AF685752

A Computer-Assisted Analysis System for Mathematical Programming Models and Solutions: A User's Guide for ANALYZE(c)

OPERATIONS RESEARCH/COMPUTER SCIENCE INTERFACE SERIES

Ramesh Sharda, Editor
Conoco/DuPont Chair of Management of Technology
Professor of Management Science & Information Systems
College of Business Administration
Oklahoma State University
Stillwater, Oklahoma, U.S.A.

A Computer-Assisted Analysis System for Mathematical Programming Models and Solutions: A User's Guide for ANALYZE[(c)]

HARVEY J. GREENBERG
Mathematics Department, Campus Box 170
University of Colorado at Denver
P.O. Box 173364
Denver, CO 80217-3364
BITNET: hgreenberg@cudenver

WITHDRAWN
IOWA STATE UNIVERSITY
LIBRARY

Kluwer Academic Publishers
Boston/Dordrecht/London

(c)This User's Guide and ANALYZE software are protected by U.S. copyright laws.

Distributors for North America:
Kluwer Academic Publishers
101 Philip Drive
Assinippi Park
Norwell, Massachusetts 02061 USA

Distributors for all other countries:
Kluwer Academic Publishers Group
Distribution Centre
Post Office Box 322
3300 AH Dordrecht, THE NETHERLANDS

Library of Congress Cataloging-in-Publication Data

Greenberg, Harvey J.
A computer-assisted analysis system for mathematical programming models and solutions : a user's guide for ANALYZE / Harvey J. Greenberg.
p. cm. -- (Operations research/computer science interface series)
"September 1992."
Includes index.
ISBN 0-7923-9322-8
1. Linear programming -- Data processing. 2. ANALYZE. I. Title. II. Series.
T57.74.G727 1993
519.7'2--dc20 92-43505
CIP

Copyright © 1993 by Kluwer Academic Publishers

All rights reserved. No part of this publication may be reproduced, stored in a retrieval system or transmitted in any form or by any means, mechanical, photo-copying, recording, or otherwise, without the prior written permission of the publisher, Kluwer Academic Publishers, 101 Philip Drive, Assinippi Park, Norwell, Massachusetts 02061.

Printed on acid-free paper.

Printed in the United States of America

Acknowledgements

I wish to thank our industrial sponsors – Amoco Oil Company, IBM, Shell Development Company, Chesapeake Decision Sciences, Ketron Management Science, GAMS Development Corporation, MathPro, Inc., and Maximal Software – for their support, including their participation in the project, itself. Earlier support was also provided by General Research Corporation and US West. Milton M. Gutterman (formerly with Amoco) provided code and testing for the TSO and SUN environments; and, Jon Jacobs (formerly with GRC), Thomas Dehne (Ketron) and David Heltne (Shell) provided code and testing for the CMS environment. Optimizers were provided by Roy Marsten (OB1) and Michael Saunders (MINOS). Support for MINOS was provided by John Stone (Ketron). John Chinneck provided code for determining an irreducible infeasible subsystem (IIS). I wish to thank colleagues on the research team – John Chinneck, Fred Glover, Weldon Lodwick, Gautam Mitra and Frederic Murphy. A special thanks goes to colleagues and students who critically reviewed earlier drafts of this User's Guide and whose comments have led to this clearer version. They are:

John Chinneck, Carleton University
Eric Dravnieks, Carleton University
Robbin Gilford, University of Colorado at Denver
Milton M. Gutterman, Amoco Oil Company
Charles Halsey, University of Colorado at Denver
Weldon Lodwick, University of Colorado at Denver
Frederic H. Murphy, Temple University

Last, but not least, an earlier version of this work was partially supported by the Office of Naval Research (contract #N-00014-88-K-0104).

Contents

Introduction

Welcome to ANALYZE, designed to provide computer assistance for analyzing linear programs and their solutions. Chapter 1 gives an overview of ANALYZE and how to install it. It also describes how to get started and how to obtain further documentation and help on-line.

Chapter 2 reviews the forms of linear programming models and describes the *syntax* of a model. One of the routine, but important, functions of ANALYZE is to enable convenient access to rows and columns in the matrix by *conditional delineation*.

Chapter 3 illustrates simple queries, like DISPLAY, LIST, and PICTURE. This chapter also introduces the SUBMAT command level to define any submatrix by an arbitrary sequence of additions, deletions and reversals. Syntactic explanations and a schema view are also illustrated.

Chapter 4 goes through some elementary exercises to demonstrate computer-assisted analysis and introduce additional conventions of the ANALYZE language. Besides simple queries, it demonstrates the INTERPRT command, which automates the analysis process and gives English explanations of results. The last 2 exercises are diagnoses of elementary infeasible instances of a particular model.

Chapter 5 progresses to some advanced uses of ANALYZE. The first is blocking to obtain macro views of the model and for finding embedded substructures, like a netform. The second is showing *rates of substitution* described by the basic equations. Then, the use of the REDUCE and BASIS commands are illustrated for a variety of applications, including solution analysis, infeasibility diagnosis, and redundancy detection.

Chapter 6 describes environment controls, such as the use of the _SETUP command. It proceeds to explain the use of various file types that provide interfaces with other systems and to offer additional computer-assisted analysis to model users. This includes the use of dictionaries, documents, execution macros, syntax files and rule files.

Chapter 7 goes through some advanced exercises. These involve more subtle uses of ANALYZE, and some practical limitations are given. It concludes with describing how ANALYZE can assist model management.

Chapter 8 describes the utilities. Of particular importance is the ability to rename rows and columns with the LPRENAME system. This allows longer than 8-character names, a limit imposed by the standard matrix file, and this provides an interface with modeling systems like GAMS, AMPL and MIMI.

Chapter 9 presents some features of the fundamental algorithms and heuristics in ANALYZE. The purpose is to enhance awareness of the meaning of the results obtained by these procedures.

Chapter 10 describes more views of a linear program. The first two are graphic plots. The first of these is an option with the BLOCK command; the second is an option with the SHOW command. The third additional view is a table that can be formed from the syntax and displayed in rectangular format with the TABLE command. Finally, schema equations are illustrated as an algebraic view of the underlying model, taken from the schema table. These views are not in the standard version of ANALYZE (see Chapter 1 on installation).

Chapter 11 extends the brief discussion of rule files, given in chapter 6, to a full presentation of what capabilities are available and how to use them. After basic terms and concepts, examples are presented from rule files that come with ANALYZE to support the INTERPRT command, which is illustrated in early chapters.

Chapter 12 applies only if you received source code (contact the author for information about availability). It explains how you can use the modules used by ANALYZE for designing your own system or changing some of the algorithms. Knowledge of FORTRAN/77 is assumed.

Appendix A gives guidance to error messages, and Appendix B describes the extended library of linear programs and how to get more information about them with suggestions for additional exercises.

Advice from a student

> Although this User's Guide is self-contained, knowledge of linear programming is assumed. The exercises in the chapters show how to use ANALYZE to retrieve information, sometimes with complex procedures, but it does not explicitly teach what information to obtain. This User's Guide is designed for people who already use linear programming, and for students who are learning linear programming. In the latter case, ANALYZE serves to support an LP laboratory, corequisite with lectures, and the exercises need classroom discussion to guide the students.

Chapter 1. Overview, Installation, and Help

Man is a tool-using animal. Without tools he is nothing, with tools he is all.
Thomas Carlyle

This chapter is designed for a quick entry into ANALYZE. The purpose of ANALYZE is to provide computer assistance for analysis. It is presumed a linear program has already been formulated, and an instance has been generated with some language. The most common input to ANALYZE is a standard (MPS) matrix file, and we shall describe alternative input forms. Besides a matrix, ANALYZE can read solution files from a variety of solvers: MINOS, MPSIII, MPSX, OB1, and OSL. Although ANALYZE does not solve the instance of a linear program, it does provide ways to analyze the linear program and its solution. The methods we shall describe comprise three levels of use.

At the very lowest level, ANALYZE provides convenient interactive query to navigate through a linear program, perhaps with a solution already obtained from some solver. A second level of use provides procedures to assist analysis in a variety of ways. Standard sensitivity questions, such as *What if...?*, *Why...?* and *Why not...?*, can be answered with easy access to information about the solution. In addition, diagnostic analysis, such as when the linear program is infeasible, can be resolved efficiently. The third level provides an artificially intelligent environment with English translations of results automatically obtained. This is especially well suited for a non-expert of linear programming.

The exercises in this User's Guide reinforce concepts learned in linear programming, but this is not a substitute for an introductory text. We freely use terms like dual price, basic solution, etc., which make more sense if one also has a text on linear programming to learn basic terms and concepts.

After describing what you have and how to begin, an overview of the ANALYZE modular design is presented to help gain an overall appreciation for what you will use to enhance your environment for analysis. These sections are brief, but they may be skimmed.

Then, in section 3, installation instructions are given for a DOS environment. You must have a 386 or 486 machine with 2mb memory. A co-processor is desirable but not necessary.

Section 4 introduces some of the conventions in the interactive language of ANALYZE, such as abbreviations and some of the more popular environmental controls. Finally, the last section tells you how to obtain on-line help and further documentation.

1.1 What you have and how to begin.

The DOS version is on one 3½" disk (1.44mb) with the following files.

File	Contents
ANALDATA.ARC	Data and Documents
ANALEXEC.ARC	Executable
FLIPDOCS.ARC	FLIP Documents
HPAGER.ARC	Pagination for HP
LPRNDATA.ARC	LPRENAME Data
LPRNEXEC.ARC	LPRENAME Executable
MOREDATA.ARC	More linear programs
PKXARC.COM	ARC extraction
VIEWDOCS.ARC	Views Documents

> Codes were compiled with Lahey(R) F77L-EM/32 5.0 and linked under DOS 3.3 (the executable uses the Phar Lap extender).

Begin by copying the original disks for a backup. To install ANALYZE on your IBM PS2 (or a clone) under DOS, you will need about 2mb on your hard disk (details for installation are given in § 1.3).

This User's Guide is designed to be read sequentially. A reference manual may be generated by ANALYZE using EXEC DOCANAL. This and other information about on-line help is given in § 1.5.

1.2 Modular design.

Figure 1-1 shows an Input/Output schematic of ANALYZE. The only required input file to ANALYZE is a *matrix file*. This is the (de facto) standard MPS file, which is accepted by all commercial optimizers and can be generated by most LP modeling languages (see §8.2 for a utility, called LPRENAME, that accepts another type of LP input file). For model analysis this may be sufficient, but other files are useful, such as a solution file from an optimizer. These (and other) files are described in Chapter 6. To get started, the data files (extracted from ANALDATA.ARC) contain a variety of examples, which are used in the exercises in Chapters 4, 5 and 7).

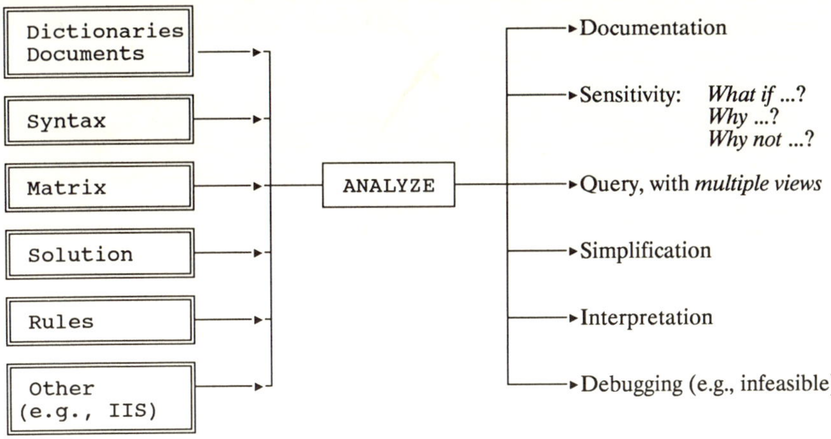

Figure 1-1. An Input/Output Overview of ANALYZE

The functions of ANALYZE are to support analysis of results and other model management tasks. The exercises in the following chapters illustrate many kinds of analysis, starting with elementary questions of interpreting optimal solutions and debugging infeasible or anomalous results. Special procedures, such as REDUCE, are described to aid model simplification; and, a multi-view architecture aids documentation as well as on-line query.

1.3 Quick install for DOS environment.

To install in a DOS environment, put the ANALYZE disk into drive a: and change directories to the path where you want to install ANALYZE. It is recommended this be an empty directory, perhaps called ANALYZE. Then, at the DOS prompt, enter:

> COPY A:INSTALL.BAT [*drive*]

Then, enter INSTALL. The default disk drive that contains ANALYZE is a:, and you can change this by entering another drive. For example, enter INSTALL B to install from drive b (note no colon after B in the drive spec).

The appropriate files will be extracted from the ARC files, and you will see a message that installation was successful. You are then ready to run ANALYZE.

Once the installation is complete, you may delete INSTALL.BAT from your hard disk. Your next step is to enter ANALYZE at the DOS prompt. After some sign-on information is displayed, you will be ready to test that you received everything and that all is well. After the ANALYZE prompt, enter EXEC ATEST. You will see a lot of messages flashing over your screen. These are from the test commands being executed. After about 10 minutes (depending upon the speed of your computer), you will be prompted to continue. Simply press Enter (or Y, followed by Enter) to complete the test. At the end of the test, you should see a message that ANALYZE appears to be working, and you are left with the ANALYZE prompt. You may then continue with some exercises. To leave ANALYZE, enter QUIT.

You can have a more comfortable screen if you enter COLORSCR before entering ANALYZE. This gives a blue background and yellow letters, but it might not work on your computer. If it does work, you can create a BAT file to execute COLORSCR before executing ANALYZE. If you have Microsoft Windows 3.0 or higher, you should look at ANALYZE.PIF and ANAL.BAT. You might need to modify them for your environment.

1.4 Some interactive language conventions.

Examples and documentation use a standard metalanguage with the following symbols (a metalanguage is a language to describe a language, in this case ANALYZE command specifications comprise the language, and their description, such as with the HELP command, uses this metalanguage, which is in fairly standard notation).

All commands and keywords are upper case. Lower case words describe a type of specification.

Square brackets [] enclose an optional specification.

Curly braces {} enclose a list of options, one of which must be specified.

An ellipsis ... denotes continued repetition of the last option.

A vertical bar | denotes exclusive options.

Example 1: `SWITCH [name [...]]`

This specifies the language syntax for the SWITCH command. The outer square brackets denote that no specification need follow the SWITCH command (in which case the switches and their values are listed). If a specification does follow, it is the name of a switch (in which case the value of the switch is toggled between ON and OFF). The inner square brackets with the ellipsis means that additional switch names can be specified. The following illustrates the possibilities.

```
SWITCH
```
...lists switches and their settings
```
SWITCH ECHO
```
...changes value of switch named ECHO
```
SWITCH ECHO ERROR
```
...changes values of two switches, named ECHO and ERROR

Example 2: `DISPLAY [ROW|COL [conditional]] [//options]`

This specifies the language syntax for the DISPLAY command. The square brackets denote options, so a valid specification is just DISPLAY with no other specification. The vertical bar that separates keywords ROW and COL means that the following specifications are valid.

```
DISPLAY ROW
DISPLAY COL
```

The `[conditional]` means another optional specification, which is a "conditional" (to be defined later) can follow the ROW or COL specification. Independent of whether the ROW or COL option is specified, there is another option indicated by `[//options]`. The double slash (//) is necessary to precede the options that can be specified. The following are valid specifications (whose meanings will be defined later).

```
DISPLAY ROW  //LEVEL
DISPLAY COL * //LEVEL
DISPLAY   //LEVEL
```

Example 3: `SOLUTION {MINOS | OB1 | OSL | XMP | MPS} [filespec]`

This specifies the language syntax for the SOLUTION command. The Curly braces indicate that one of the options must be specified, and the vertical bars indicate that exactly one must be specified. The square brackets indicate that the file specification (filespec) is optional. This means the following specifications are valid.

```
SOLUTION MINOS
SOLUTION OB1
SOLUTION OSL
SOLUTION XMP
SOLUTION MPS
SOLUTION MINOS MYFILE
```

In all but the last case, a file is not specified. In the last case, the file is specified as MYFILE.

Example 4: `SCHEMA [DISPLAY [options]|CLEAR|{LOAD | SAVE} [filespec]]`

This specifies the language for the SCHEMA command. The outer square brackets mean that just SCHEMA can be specified with no further specification (as before, the meaning is described later; here only the metalanguage is illustrated). What follows means the following are valid.

```
SCHEMA DISPLAY [options]
SCHEMA CLEAR
SCHEMA LOAD [filespec]
SCHEMA SAVE [filespec]
```

Note the use of the curly braces for the LOAD and SAVE specifications following SCHEMA. Whichever of those are specified, a filespec (file specification) is optional. There are also options following DISPLAY, and the square brackets state that SCHEMA DISPLAY is a valid specification without options.

As we go through the exercises in subsequent chapters, you will experience some conventions about the interactive language (FLIP). One is that you must LOCK CAPS, because the interactive command language is case-sensitive, and all commands and keywords use capital letters.

Telescopic abbreviations are allowed as long as there is no ambiguity. For example, the telescopic abbreviations of DISPLAY are:

```
D
DI
DIS
DISP
DISPL
DISPLA
```

Since no other command begins with D, this is an allowed abbreviation. On the other hand, the READIN and RETURN commands both begin with RE, so at least 3 keystrokes are required for each of those commands.

Another convention is allowing *string variables*, which are local to the session, to reduce the number of keystrokes for repeated specifications. For example, one may define the string `R` as:

```
STRING R = READIN MATRIX
```

then use 'R whenever you want to read in a matrix file. The single quote (') is the reserved character for denoting a string variable.

The STRING command can also be used to clear a string by specifying STRING *name*. To clear all strings, specify the asterisk (*) for the *name*.

It is also useful to change the width and length of your screen. This is done with the SCREEN command. The default (upon installation) sets WIDTH=79 and LENGTH=23. You can see your settings anytime by entering SCREEN (or simply SCR) at the ANALYZE prompt. To change either or both of these simply enter the desired number(s), such as SCR L=9999 to make the length virtually infinite. The screen setting is also used when spooling output to a file to determine page ejection (see OUTPUT command).

The strings, screen and other settings may be set in your PROFILE, which is an execution file that is done before beginning the interactive session. This is described in Chapter 6.

> Additional information is available on-line (see next section). Especially look at the help for each command, and PRINT QUICKREF.DOC for a quick reference to all commands (the .DOC suffix is not needed if you are printing from within ANALYZE).

1.5 On-line documentation.

ANALYZE has a variety of facilities for providing on-line documentation. At the ANALYZE prompt you may enter ? to obtain a general introduction. The list of commands is shown by entering HELP to the ANALYZE prompt. For information about any particular command, enter the command after HELP. For information about all commands, such as when generating your own file for printing, enter HELP *.

> It is recommended you try the HELP command
> (after EXEC ATEST is successful).

There is a dictionary that will help guide you to the nomenclature and other documentation sources. Enter ? DIR at the ANALYZE prompt to see what directory files are available, which you may PRINT. In particular, note the DIRDOCS dictionary entry, which guides you to some documentation files that you could print for more explanations (see §6.2 to create your own dictionaries).

It is recommended that you enter EXEC DOCANAL at the ANALYZE prompt. This will generate a file, called ANALYZE.DOC, that contains most of the help and documentation available on-line. You might also examine the file DOCANAL.EXC (by printing it), which generates ANALYZE.DOC. This will suggest how you might make your own, personalized documentation.

Each PRINT uses the screen width and length settings, even when the output is a file (specified by the OUTPUT command). When you EXEC DOCANAL, you will see the screen settings and a prompt is given (from the _PAUSE command in DOCANAL.EXC) to see if you want to continue. If you forgot to increase the screen LENGTH, enter N to abort the execution of the documentation and specify SCREEN L=*length* to set the screen length to some integer (*length*), like 60, depending on your printer (see §6.3 for how to have this set with a PROFILE execution file before beginning your session).

When printing to a file, as above, page ejection (from the _EJECT command) is merely the FORTRAN 1H1 format. The output file, therefore, has a line with a 1 in column 1 wherever there should be a new page, which is determined by the SCREEN LENGTH setting (see §8.1 for automatic conversion for an HP Laser Jet printer).

Once you become familiar with how ANALYZE works, you may add your own documentation for others to have. Dictionaries and Documentation files are two ways to supplement the on-line documentation, and the DOCANAL execution file is one way to obtain hard copy. After reading Chapter 6, you will know more ways to make ANALYZE suit your environment.

Chapter 2. Anatomy of a Linear Program

Modeling is like carpentry —— we sometimes hit the nail right on the thumb.
Mark Twain (paraphrase)

In this chapter we consider the parts that comprise a linear program and how to use them for effective delineation of rows and columns you might want to see. We begin with the algebraic form, which views the model as a collection of variables related by a system of equations and inequalities with an objective to minimize or maximize some linear form. Beyond the algebra, linear programs have composition rules that comprise its syntax. This is presented with its role in supporting analysis. Finally, the methods of row and column delineation are presented with examples.

2.1 Algebraic form.

One way to express an LP model, which is the way of most textbooks, is first to define sets, domains, data tables and variables, and second to define the constraints. To illustrate, here is a production-distribution model.

Given: 1. A collection of plants distinguished by their location and processes of operation. Each process requires raw material input and produces finished products.

2. Markets, distinguished by location and product.

3. Transportation links from plants to markets.

Data: a_{ipj} = unit amount of raw material i used by process p at plant j;

α_{pjk} = unit yield of product k from process p at plant j;

S_i = total supply of raw material i;

c_{pj} = unit operation cost of process p at plant j;

C_j = capacity limit of plant j, measured in terms of its total output;

s_{jmk} = unit shipping cost of product k from plant j to market m;

D_{mk} = demand for product k required to be satisfied in market m.

Conditions: If there is no link from plant j to market m for product k, the value of $s_{jmk} = \infty$.

Model: Let P_{pj} = level of production using process p at plant j;

Let T_{jmk} = amount of product k sent from plant j to market m.

Let COST = total cost of production and transportation.

Minimize COST subject to: $P_{pj}, T_{jmk} \geq 0$;

COST =	$\Sigma_{pj} c_{pj} P_{pj} + \Sigma_{jmk} s_{jmk} T_{jmk}$	
A(i) =	$\Sigma_{pj} a_{ipj} P_{pj} \leq S_i$	(Raw material availabilities)
C(j) =	$\Sigma_p P_{pj} \leq C_j$	(Capacity limits)
B(j,k) =	$\Sigma_p \alpha_{pjk} P_{pj} - \Sigma_m T_{jmk} = 0$	(Balance equations)
D(m,k) =	$\Sigma_j T_{jmk} \geq D_{mk}$	(Demands)

Notice that we began our expression with *Let* ... When we express the *model*, we write the objective and the constraints. This is the *algebraic form*. Alternatively, we could express the model in *schema form*, as shown in figure 2-1.

	P(p,j)	T(j,m,k)	
COST	PCOST	SHIPCOST	...MIN
A(i)	CONSUME		<= SUPPLY
B(j,k)	YIELD	-1	= 0
C(j)	1		<= CAPACITY
D(m,k)		1	>= DEMAND

Figure 2-1. A Schema of the Production-Distribution Model

The rows have been classified with COST the objective row, and the other row types begin with A, B, C and D, respectively. The activities have been classified as Production and Transportation. Tables PCOST, SHIPCOST, CONSUME, SUPPLY, YIELD, CAPACITY and DEMAND had to be declared, and they correspond, respectively, to the given c, s, a, S, α, C, and D. The subscripts in the original problem definition become *domains* in this expression (and in algebraic modeling languages). Each domain is a cross product of sets. The sets are:

i = raw material;
p = process;
j = plant (location);
k = finished product;
m = market location.

When forming the name of a row or column, its type is distinguished by its first character. Then, its domain is used (without the parentheses and commas).

For example, consider the transportation activity T(j,m,k) for the particular plant j=DE (a code for Denver), the particular market m=CH (a code for Chicago), and the particular finished product k=TB (a code for table). Then, the column name is TDECHTB. This syntax will be formally described in the next section. Before doing so, we define the algebraic form assumed by ANALYZE.

The LP *canonical algebraic form* is defined as:

$$\text{LP: } \textit{optimize} \text{ CX: } Y = AX, L \leq (X,Y) \leq U,$$

where *optimize* can be MINIMIZE or MAXIMIZE. The data for an instance of the linear program (LP) are:

C...............an N-vector, called the *objective coefficients*;
A...............an M×N array, called the *matrix*;
L...............an (M+N)-vector, called the *lower bounds*;
U...............an (M+N)-vector, called the *upper bounds*.

M is the number of rows, and N is the number of columns.

The solution values in LP are:

X...............*activity levels* associated with columns (N-vector);
Y...............*levels* associated with rows (M-vector);
P...............*prices* associated with rows (M-vector);
D...............*reduced costs* associated with columns (N-vector).

The prices and reduced costs are solutions to the *dual* of LP, which we shall not describe here.

We sometimes refer to (C, L, U, X, Y, P, D) as the *rim* of LP, and the nonzero coefficients in A as the *body*. The body embodies relations among activities, whereas each rim element is associated with a single row or column.

Usually, one does not formulate an LP with homogeneous equations (AX−Y = 0). Instead, we typically see:

$$Ax \leq b \quad \text{or} \quad Ax \geq b,$$

where b is called a *right-hand side* vector, sometimes abbreviated RHS. ANALYZE transforms such expressions by defining X=x and Y=AX. Then, in the first case, the lower bound on Y is −∞ and the upper bound is b; in the second case, the lower bound on Y is b, and the upper bound is ∞.

For the dual prices to represent economic marginal prices, as for supply and demand constraints, the LP must be equivalent to the second form for minimization (i.e., $Ax \geq b$) and the first form for maximization (i.e., $Ax \leq b$). This is not necessary to use ANALYZE, but the interpretations must be made with care if the canonical form is not used. In particular, automatic interpretation is not guaranteed if the model has equations, such as balances, that are not equivalent to the canonical form. One can, for example, model a standard transportation problem as:

$$\text{Minimize } \Sigma_{ij}\, c_{ij}\, x_{ij}\text{: } x \geq 0,$$
$$\Sigma_j\, x_{ij} \quad = \quad s_i \qquad \text{for all } i$$
$$\Sigma_i\, x_{ij} \quad = \quad d_j \qquad \text{for all } j.$$

What makes this non-canonical is that the supply equations need to be multiplied by −1. Then, and only then, is this equivalent to the canonical form:

$$\text{Minimize } \Sigma_{ij}\, c_{ij} x_{ij}\text{: } x \geq 0,$$
$$-\Sigma_j\, x_{ij} \quad \geq \quad -s_i \qquad \text{for all } i$$
$$\Sigma_i\, x_{ij} \quad \geq \quad d_j \qquad \text{for all } j.$$

In the following chapters you will learn how to see many views of the instance of the linear programming model, including the algebraic and block schema views. Construction of many of the views depends on the model's syntax, which we consider next.

2.2 Model syntax.

The *syntax* of a model is its rules of composition. This was introduced in the previous section and illustrated with a production-distribution model. We see that the name of a row or column has information about its meaning.

Suppose, for example, we specify the following set elements.

i = {S,W} : S means steel, W means wood;
p = {1,2,3} : 1 means process 1, 2 means process 2, 3 means process 3;
j = {N,S} : N means North and S means South;
m = {DE,CH} : DE means Denver and CH means Chicago;
k = {C,T} : C means chair and T means table.

Suppose further that the shipping links are only those shown in figure 2-2. Then, the matrix for this instance is pictured in figure 2-3. Note that the columns are printed vertically, which is how ANALYZE pictures matrices. Only the signs of the nonzero coefficients and right-hand sides are shown.

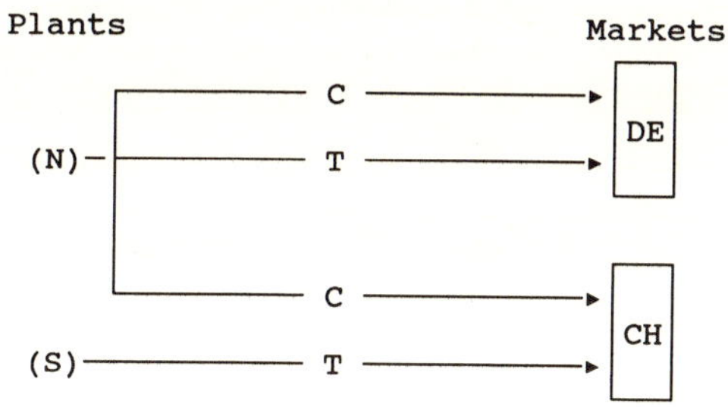

Figure 2-2. Shipping Links for An Instance of the Production-Distribution Model (Arc Labels are Products)

	P1N	P1S	P2N	P2S	P3N	P3S	TNCHC	TNDEC	TNDET	TSCHT		
AS	+	+			+	+					<=	+
AW			+	+	+	+					<=	+
BNC	+				+		-	-			=	0
BNT			+		+				-		=	0
BST				+		+				-	=	0
CN	+		+		+						<=	+
CS		+		+		+					<=	+
DCHC							+				>=	+
DCHT										+	>=	+
DDEC								+			>=	+
DDET									+		>=	+

Figure 2-3. A Qualitative View of the LP Matrix for an Instance of the Production-Distribution Model

For example, consider the first column, P1N. This is the name of activity P(p,j) for p=1 and j=N. It uses steel, indicated by the + entry in row AS, which is the availability constraint, A(i) for i=S. The activities that produce with process 2, namely P2N and P2S, each use wood; and, those with process 3 use fixed shares of steel and wood. The capacity row, CN, limits the total capacities used by the associated production activities, P1N, P2N and P3N.

When we refer to the model syntax in ANALYZE, we generally mean the rules by which the names were formed - that is, the *handle*. A summary of the LP Anatomy is given in figure 2-4.

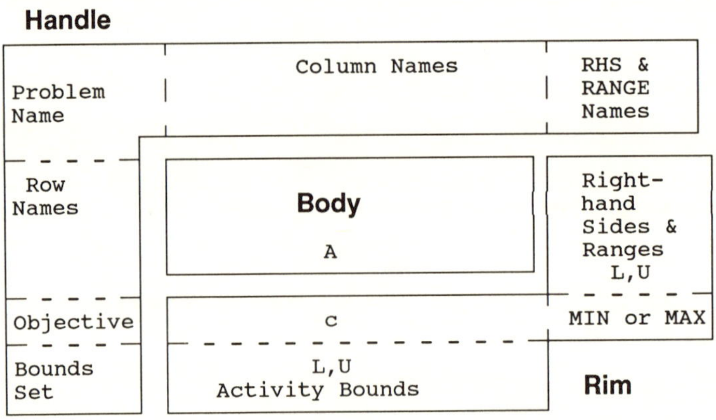

Figure 2-4. Anatomy of An LP

The handle is what we call things, notably the names of the rows and columns. In some modeling environments multiple sets of right-hand sides and bounds are used to define different scenarios of the same model (but different data). The rim sits on the edge with numerical values that are attributes of individual rows and columns, and the body represents relations among activities.

2.3 Conditional delineation of rows and columns.

The names and rim values are used to delineate rows and columns for query and other functions of ANALYZE. Here we give a brief, isolated introduction; the next chapter illustrates this with queries for a specific LP.

A *conditional* is a specification of row or column attributes:

Name................ string with no embedded blanks or other delimiters: { ,=()};
Nonzeroes......... range for its number of nonzero coefficients;
Status................ list from {L U B I};
Rim relation..... rim_key relation rim_key;
Rim range......... rim_key range.

The name string can contain 1 or more *mask characters*, where the mask character is the asterisk (*). Trailing blanks are treated as mask characters. For example, the strings ABC and ABC** will match the same set of names. Except for trailing blanks, a mask character is matched by any single character in that position. For example, A*C is matched by ABC, AXC, and by ABCDEF.

In addition to the usual name_mask, one can specify a *sliding name mask*. In the usual case, the positions are fixed. For example, the name ABC matches the mask A*C, but the name XABC does not. The name_mask specification, A*C, says that the name must have A in position 1 and C in position 3. A sliding name mask is specified by using the double quote (") as a special character (which can therefore not be in the names). The sliding mask, "A*C, is matched by names, ABC, XABC and XXXAYC. The match is based on the name matching the substring A*C (with * still a wild character) anywhere, not just in the first 3 positions.

The general mask combines the fixed and sliding masks with the form: [fixed]["sliding]. If " does not appear, the mask is fixed. If " is the first character, the mask is sliding. More generally, a name matches if first it matches the fixed mask, then the remaining substring matches the sliding mask.

Examples of name masks:
  Ordinary masks (fixed positions):
    *A*B..............character 2 must be A and character 4 must be B;
    ABC................name begins with ABC;
    ***AB............name has AB as characters 4 and 5;
    *.....................any name qualifies.
  Sliding masks (indicated by " as first character):

```
    "*A*B......name contains a substring with A in its 2nd
                  position and B in its 4th position
    "ABC.......name contains ABC as a substring somewhere
    "***AB.....name contains a substring of length 5 whose
                  last 2 characters are AB
  General masks
    A"BC.......name begins with A and contains BC after the
                  1st character
    ***"AB.....name contains AB after the 3rd character
```

A range specification, such as for delineating the number of nonzeroes, is of the form *v*[/*v*], where *v* denotes a value. If the second value is omitted, it is treated as equal to the first. For example, N=1 and N=1/1 both specify that there is to be exactly 1 nonzero. Further, the specification N=*range* abbreviates NONZERO=*range*, where NONZERO is a keyword for the rim object: number of nonzero coefficients.

In general, when the range specification has two values, v_1/v_2, the following parsing is done. If $v_2 \geq v_1$, the range is the interval, $[v_1, v_2]$ if both v_1 and v_2 are finite. If v_1 is –* (minus infinity) or v_2 is * (infinity), the interval is open at the appropriate end(s). If, on the other hand, $v_2 < v_1$, the range is the interval, $[v_1, v_1 + v_2]$ if $v_2 > 0$, and $[v_1 + v_2, v_1]$ if $v_2 < 0$. For example, the interval [10,12] can be expressed by 10/12, 10/2 and by 12/-2.

Examples of Nonzero ranges (N abbreviates NONZERO):
- `N=0`....................null (no nonzeroes);
- `N=1`....................singletons (exactly 1 nonzero);
- `N=0/1` ....either null or singleton;
- `N=1/5` ....between 1 to 5 nonzeroes (inclusive);
- `N=1/*` ....1 or more nonzeroes (* denotes infinity);
- `N=10/2`..............10 to 12 nonzeroes;
- `N=10/-2`............8 to 10 nonzeroes.

Another keyword is STATUS. This is the *solution status*, where

- `L`......... nonbasic at Lower bound;
- `U`......... nonbasic at Upper bound;
- `B`......... Basic (and feasible);
- `I`......... Infeasible (and basic).

For example, `STATUS=L` (or simply `S=L`) specifies that the solution status must be L (nonbasic at lower bound); `S=LU` specifies that the solution status must be nonbasic (either L or U). Any combination of the 4 status values, L, U, B, and I, can be specified contiguously (no embedded blanks), and the delineation is interpreted as an inclusive-or.

Examples of Status specifications (S abbreviates STATUS):
- `S=B`.....................Basic and feasible;
- `S=I`.....................Infeasible (basic);
- `S=BI` ....Basic (B or I);
- `S=LU` ....Nonbasic (at Lower or at Upper bound)
- `S=LUB` ....Feasible (L or U or B).

The other keywords for rim delineation are called *rim keys*. The rim keys are:

L .... Lower bound;
U .... Upper bound;
C .... Objective coefficient (sometimes called *cost*);
X .... Activity level for column;
Y .... Level for row (Y=AX);
P .... Price for row;
D .... Price (or reduced cost) for column (D=C-PA).

Rim relations specify two rim keys separated by one of the relations: {LT, LE, EQ, GE, GT, NE} with the following meaning.

LT .... less than;
LE .... less than or equal to;
EQ .... equal to;
GE .... greater than or equal to;
GT .... greater than;
NE .... not equal to.

For example, the condition X EQ L means the column's activity level must equal its lower bound.

A rim range is a numeric range that the rim element's value of the row or column must satisfy. For example, P=0 means the row price must be zero, and X=1/* means the column activity level must be at least 1 (where * denotes infinity in a numeric range specification). The following examples are in CONDNL.DOC, which is available on-line.

Examples of Rim Range specifications:
X=0 .... column activity level equals zero;
Y=1 .... row level equals 1 (where Y=AX);
L=-* .... lower bound equals negative infinity;
U= * .... upper bound equals infinity;
C=0/* .... cost coefficient is non-negative;
D=-*/-1 .... column reduced cost does not exceed –1.

Examples of Rim Relation specifications:
L EQ U .... lower bound equals upper bound;
X LT U .... column activity level is strictly less than its upper bound;
P GT C .... row price is strictly greater than its cost (which is 0);
Y NE L .... row level not equal to its lower bound;
D EQ C .... column reduced cost equals its cost.

The specifications can be compounded to form a conditional. Here are some examples:

`* S=LU D=0`......................activity with nonbasic solution status and zero reduced cost;

`P X EQ L D=1/*`.............activity whose name begins with P, whose level is equal to its lower bound, and reduced cost in the range [1,∞);

`*AB S=B X EQ U`.............basic activity name has AB in its 2nd and 3rd characters, and whose level equals its upper bound;

`* S=B N=1`........................basic activity with 1 nonzero coefficient.

In a conditional specification, the name must appear first. Then, the keyworded specifications can appear in any order.

Chapter 3. Simple Queries

The pure and simple truth is rarely pure and never simple.
Oscar Wilde

In this chapter we start with the sign-on screen and describe some of the interactive conventions, such as abbreviations of commands and keyword parameters. Then, we shall illustrate how to query model and solution values. The dialogue is designed for you to follow the exercises, and it is recommended you do so. Change to the directory where you installed ANALYZE. Then, at the DOS prompt enter:

> ANALYZE

3.1 The sign-on and basic interactive conventions.

When you first enter ANALYZE, credits will appear, as shown in figure 3-1. This is simply printing the file CREDITS.DOC, which you may delete or rename to avoid printing in subsequent sessions. You answer the prompt simply by pressing Enter (Y is the default response).

```
                              CREDITS

ANALYZE was written by Dr. Harvey J. Greenberg, University of
Colorado at Denver.  The development was sponsored by a consortium
of companies as part of a project to develop an "Intelligent
Mathematical Programming System." The companies are:  Amoco Oil
Company, IBM, Shell Development Company, Chesapeake Decision Sciences,
GAMS Development Corp., Ketron Management Science, MathPro Inc., and
Maximal Software.  Partial support was also provided by the Office of
Naval Research (contract #N-00014-88-K-0104).

Some programming and testing was provided by Thomas Dehne (Ketron),
Milton M. Gutterman (Amoco), Jon Jacobs (GRC), and David Heltne
(Shell).  James Kalan gave advice on some of the numerical methods.

Dr. Greenberg is responsible for any errors in the programs but
assumes no liability, as this was written as part of an
"experimenter's workbench" to aid research in mathematical
programming.
```

Figure 3-1. Initial Screen gives Credits

The Version shown in figure 3-2 may not agree with your version. This may mean you should obtain the latest, but the rest of this User's Guide should still be a valid introduction to ANALYZE. Note the message to PLEASE LOCK CAPS; ANALYZE is case-sensitive, and all commands and keywords require capital letters. The names of rows and columns in the model, however, may contain lower-case letters, and they must be entered as such.

```
                    ANALYZE    Version 10.0
                               by
                     Harvey J. Greenberg

                : Enter  ?  for introduction :
                : Enter HELP to see commands :
                :............................:


 PLEASE LOCK CAPS

                ... Enter  ? ? for help with dictionary
ANALYZE ... H
        ANALYZE  has  16 session control commands:
    EXECUTE  HELP     OUTPUT   PRINT    QUIT     RETURN
    SCREEN   STRING   SWITCH   RECAP    _EJECT   _PAUSE
    _SETUP   _DICTNRY _KEYWORD _TYPE
        and  26 procedural commands:
    ADDRIM   AGGREGAT BASIS    BLOCK    COUNT    DISPLAY
    EXPLAIN  FREE     GRAPH    INTERPRT LIST     PICTURE
    RATEOF   READIN   REDUCE   RENAME   SCHEMA   SHOW
    SOLUTION SUBMAT   SUMMARY  SYNTAX   TABLE    TALLY
    TRACE    WRITEOUT

   Enter HELP * to get help for each command
```

Figure 3-2. ANALYZE Sign-on and Command List (Response to HELP)

After the initial sign-on, you may enter ? for a general introduction; and, you may enter HELP (or simply H) to see the commands. If you do the latter, you will see the list of commands, as shown in figure 3-2, and you may enter any command after HELP for on-line help.

A Discrepancy

> On your PC the ANALYZE ... prompt appears on a line by itself, but figure 3-2 shows my response (H) on the same line. This appears throughout this User's Guide to save the extra line because some figures are rather long. Such edits have been minimized, and what you see in the User's Guide should agree with what you see on your screen. There are a few instances where some minor editing was done to accommodate the width of the User's Guide page, which is less than the width of your screen.

Among the conventions you will experience, ANALYZE (more accurately, the FORTRAN **L**anguage **I**nteractive **P**rocessor) allows telescopic abbreviations. The reason that H means HELP is that there is no other command beginning with H. Other telescopic abbreviations of HELP are HE and HEL. This carries to keywords and parameters within commands. For example, to read a matrix file into memory, the complete command specification is:

ANALYZE ...READIN MATRIX *filespec*

We may, however, enter:

ANALYZE ...REA M *filespec*

The least abbreviation of READIN is REA because RETURN is another command that starts with RE. We may abbreviate the keyword (file type) MATRIX as M, MA, MAT, MATR or MATRI. The *filespec* is a file name, but it may have a prefix (i.e., a path) or a suffix other than what has been established in the _SETUP command. For this introduction, we shall bypass file naming conventions and simply use the file name for any *filespec* we need. Later, in chapter 6, we shall describe how to use the _SETUP command before and during a session to reset the default prefixes and suffixes of the various file types.

The commands are grouped functionally into two kinds: *session control* and *procedural*. The session control commands fulfill functions common to any interactive session, regardless of application; procedural commands are specific to ANALYZE functions. Most of the session control commands will enter our exercises naturally, so let's immediately begin to apply the procedural commands that provide basic query functions.

3.2 Basic query commands.

To begin, we read an LP into memory, called WOODNET. Its matrix file was read in when you tested ANALYZE (by EXEC ATEST), followed by a MINOS solution file; then, a binary file of the resident LP matrix and solution, called a PACKED file, was saved (with the WRITEOUT command). We begin by reading this packed file into memory, as shown in figure 3-3.

Advice from a student

> From this point, through subsequent chapters, we shall examine models and data from files that came with ANALYZE. You might find it useful (though not necessary) to have a printed copy of these files. A directory of input files, partitioned by type, appears in DIRDATA.DOC. Print these, if you wish, using the DOS print command, and keep them handy in alphabetical order. (Further information about these file types are postponed until chapter 6, but their meanings will become clear as we go through examples.)

Do not worry about the READIN and other input/output commands yet. We shall cover that later. For now, we want to concentrate on how to navigate through a model instance (i.e., a matrix) and its solution. Examine figure 3-3 after the REA P WOODNET, and you see that SUM was entered. (Try it yourself.) This is an abbreviation for the SUMMARY command, which gives some statistics about WOODNET.

```
ANALYZE ... REA P WOODNET
 Packed LP read from WOODNET.PCK

ANALYZE ... SUM
 Problem WOODNET is to MINIMIZE COST.
 RHS=HOMOGEN  RANGE=none  BOUND=BNDSET
 Source file: WOODNET.PCK
 Solution: MINOS    Solution Status: OPTIMAL
 LP Statistics:
     36 Rows (1 free) and 93 Columns (20 fixed)
     221 Nonzeroes (23 distinct... 148 ones).
     Density = 6.60096%  Entropy = 10.4072%  Unity = 66.9683%
 Row    SPIDE    has the most nonzeroes = 6.
 Column TMOSECH  has the most nonzeroes = 3.
```

Figure 3-3. Reading WOODNET Into Memory

The first part of the summary gives the problem name, which need not be the file name we specified, especially if we have many instances - that is, scenarios. It then tells us the objective is to minimize COST (which is one of the rows in the matrix). The RHS, RANGE and BOUND names are those that were read in from the matrix file. The source file is the packed file that we just read in, and no syntax file has been read (we'll get to that later). The solver was MINOS, and the resident solution is optimal.

Now consider the LP Statistics that are shown next in the SUMMARY output in figure 3-3. Among the 36 rows, one is *free*. This is the objective row, named COST. (Every LP must have at least one free row, which is an equation for which there is no constraint on the row level – that is, the row variable has infinite limits, whereas non-free rows have lower and/or upper limits on their levels.) Of the 93 columns, 20 have *fixed* activity levels - that is, where its lower bound (L) equals its upper bound (U). There are 221 nonzeroes in the matrix, not counting bounds and right-hand sides (i.e., from the COLUMNS segment of a standard matrix file, which includes the objective coefficients). Of these, there are only 23 distinct values (magnitudes only), and 148 of the nonzeroes are ones (±1).

The *Density* is the percentage of nonzeroes in the matrix relative to its maximum possible, which is the product of the number of rows and number of columns. In WOODNET we have 221 nonzeroes, which is about 6.60096% of the maximum, 36×93. The *Entropy* is the percentage of distinct values relative to its maximum possible, which is the number of nonzeroes. In WOODNET we have 23 distinct values, which is about 10.4072% of the maximum possible, 221. The percentage of *Unity* values is computed by 148/221, which is about 66.968%.

It is sometimes useful to know how many nonzeroes are in a row or column. The summary gives a row and a column having the greatest numbers, where free rows, like COST, are excluded. Other statistics may be given, which we illustrate later.

To prepare for some analysis, we display the demand rows, which all begin with D. In figure 3-4 we abbreviate DISPLAY with D and ROW with R. Thus,

```
D R D
```

means DISPLAY all ROWS whose name begins with D. If we forgot the syntax of the DISPLAY command, or if we want to see some examples on-line, we could enter:

```
H D
```

This gives help for the DISPLAY command.

ANALYZE responds with the display, as shown. (You may see a prompt, not shown in figure 3-4, asking if you want to continue before you see the last row (`DWASF`).) In the next chapter we analyze the meaning of the dual prices, which may be regarded as consumer prices. One thing to notice is the way WOODNET was formulated. All the demand rows are homogeneous equations – that is, we see that each lower bound = 0, and each upper bound = 0. Instead of putting demand requirements as right-hand sides, they appear as bounds on associated consumption activities.

```
ANALYZE ... D R D
 20 Rows are in submatrix.
 ROW       STAT  LEVEL (Y)     LO_BOUND       UP_BOUND       PRICE (P)
 ---------------------------------------------------------------------
 DMOCH      L          0             0              0           73.000
 DMODE      L          0             0              0           75.000
 DMOLA      L          0             0              0           68.000
 DMOSE      L          0             0              0           71.000
 DMOSF      L          0             0              0           65.000
 DOKCH      L          0             0              0           60.000
 DOKDE      L          0             0              0           69.000
 DOKLA      L          0             0              0           62.000
 DOKSE      L          0             0              0           65.000
 DOKSF      L          0             0              0           60.000
 DPICH      L          0             0              0           22.000
 DPIDE      L          0             0              0           22.000
 DPILA      L          0             0              0           17.000
 DPISE      L          0             0              0           20.000
 DPISF      L          0             0              0           24.000
 DWACH      L          0             0              0           78.000
 DWADE      L          0             0              0           68.000
 DWALA      L          0             0              0           75.000
 DWASE      L          0             0              0           78.000
 DWASF      L          0             0              0           70.000
```

Figure 3-4. Displaying the Demand Rows of WOODNET

In general, the DISPLAY command may be followed by the specification of ROW or COLUMN (or simply R or C, respectively), then by a conditional, namely those rows whose name begins with D. This creates the matching rows in the *working submatrix*. Initially, after reading in the LP, the submatrix has no rows or columns. Whenever we issue a command with a conditional delineation, the rows and/or columns delineated comprise the working submatrix.

If we do not want to change the working submatrix, we omit the conditional. Thus, if we would now specify DISPLAY (or simply D), we would see the same output as if we specified the conditional that defines the same submatrix. This feature is global, in that some other command could cause the creation of a submatrix, yet DISPLAY will perform the same. Only when a conditional is specified, does the submatrix get re-defined.

Figure 3-5 continues with a column display. The specification means the following.

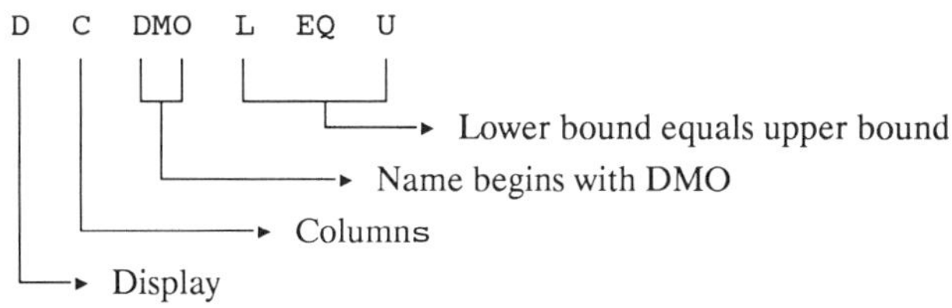


```
ANALYZE ...D C DMO L EQ U
 5 Columns are in submatrix.
 COL       STAT  LEVEL (X)     LO_BOUND      UP_BOUND      PRICE (D)
 --------------------------------------------------------------------
 DMOCH1     L        25.000        25.000        25.000        73.000
 DMODE1     L        20.000        20.000        20.000        75.000
 DMOLA1     L        20.000        20.000        20.000        68.000
 DMOSE1     L        10.000        10.000        10.000        71.000
 DMOSF1     L        30.000        30.000        30.000        65.000
```

Figure 3-5. A Display of Some Columns

Notice that the notation EQ was used for the equality relation between two rim elements (L and U). If a numeric range is desired, the equality symbol – is used; for example, L = 25 delineates columns whose lower bound equals 25, and L=20/30 delineates columns whose lower bound is in the interval, [20, 30].

You may also choose which of the five fields are displayed by specifying a parameter option, delimited by two forward slashes (//). Try each of the following specifications and compare with figure 3-5.

```
...D //          <------ just the 5 column names appear
...D //ST        <------ just the STAT field appears
...D //U         <------ just the UP_BOUND field appears
...D //LEV,P     <------ just the LEVEL and PRICE appear
```

Figure 3-6 shows a PICTURE and a LIST of the nonzeroes in the submatrix. Note that the submatrix has changed: not all Demand rows are in the submatrix. This is because the PICTURE command removes null rows and columns before picturing the submatrix. In this case we specified fixed columns (i.e., whose levels are fixed) whose names begin with DMO.

```
ANALYZE ...PIC
                                Picture of submatrix
         D D D D D
         M M M M M
         0 0 0 0 0
         C D L S S
         H E A E F
         1 1 1 1 1
 DMOCH   -           = 0
 DMODE     -         = 0
 DMOLA       -       = 0
 DMOSE         -     = 0
 DMOSF           -   = 0
ANALYZE ...L
      0  = DMOCH    =- DMOCH1
      0  = DMODE    =- DMODE1
      0  = DMOLA    =- DMOLA1
      0  = DMOSE    =- DMOSE1
      0  = DMOSF    =- DMOSF1
```

Figure 3-6. A Picture and List of the Submatrix

The default form of the list output is an equation listing, which is an algebraic view - that is, the equations for the submatrix. We can obtain other forms, such as the tabular listing in figure 3-7. The use of the COLUMN parameter (delimited by //) lists the nonzeroes sorted by columns. (Try H L to get help with the LIST command, which describes the forms of output.) The equations, in figure 3-6, show most clearly that each fixed demand column is really a right-hand side. (A reason for this homogeneous formulation, which puts limits and requirements as activity bounds, is motivated by flow tracing, which we shall introduce in the next chapter.) The picture offers a qualitative view of the nonzero pattern in the submatrix, while any form of the LIST gives the numerical values.

```
ANALYZE ...L //COL
  DMOCH1      DMOCH       -1.0
  DMODE1      DMODE       -1.0
  DMOLA1      DMOLA       -1.0
  DMOSE1      DMOSE       -1.0
  DMOSF1      DMOSF       -1.0
```

Figure 3-7. An Tabular Listing of the Submatrix with the COLUMN Parameter

The conditionals used in the query commands do not allow more complex submatrix construction, such as logical-or delineation. For that purpose we enter the SUBMAT *sub-command level.* As shown in figure 3-8, we first clear the submatrix - that is, make it null - by specifying CLEAR (we could have abbreviated CL). Then, we perform a series of additions and deletions to arrive at the submatrix pictured.

That is, after entering the SUBMAT subcommand level, we entered:

```
A R *          ◂—— Add all rows
A C S U=*      ◂—— Add columns whose name begins with S
                   and whose upper bound is infinite (indicated by *)
D R COST       ◂—— Delete the COST row
               ◂—— blank line to leave SUBMAT command level
```

```
ANALYZE ...SUBMAT CLEAR
 Submatrix has  0 rows,  0 columns, and 0 nonzeroes
 SUBMAT (enter ? for menu)...?
 SUBMAT subcommands:
    For delineation:
 ADD | DELETE | REVERSE   {ROW|COL} [conditional]
    For later EXECUTE:  SAVE  [filespec]
 ...Enter RETURN to return to main command level
 SUBMAT (enter ? for menu)...A R *
 SUBMAT (enter ? for menu)...A C S U=*
 SUBMAT (enter ? for menu)...D R COST
      (this line is blank...a return was entered from the keyboard)
ANALYZE ...PIC
                               Picture of submatrix
          S S S S
          M O P W
          O K I A
          S S S S
          E E F E
          1 1 2 1
 SMOSE    +       =  0
 SOKSE      +     =  0
 SPISF        +   =  0
 SWASE          + =  0
```

Figure 3-8. The SUBMAT Sub-Command Level

In figure 3-9 we re-enter SUBMAT, but this time we do not clear the submatrix, and we are reminded it has 4 rows and 4 columns (c.f., figure 3-8). We then proceed to add the COST row and the columns whose name begins with T and whose activity level (X) is strictly greater than its lower bound (L). We press Enter to return to the main ANALYZE command level, then we specify an equation listing of the submatrix.

```
ANALYZE ...SUB
 Submatrix has  4 rows,  4 columns, and  4 nonzeroes
 SUBMAT (enter ? for menu)...A R COST
 SUBMAT (enter ? for menu)...A C T X GT L
      (this line is blank...a return was entered from the keyboard)
ANALYZE ...L
    MIN    COST   = + 55 SMOSE1 +  50 SOKSE1 +  12 SPISF2 + 60 SWASE1
                    + 18 TMOSECH +  20 TMOSEDE +  10 TMOSESF
                    +  5 TMOSFLA + 8 TMOSFSE  + 11 TOKCHDE
                    + 10 TOKDECH +  15 TOKDESE +  5 TOKLASF
                    + 10 TOKSESF +  12 TOKSFDE +  5 TOKSFLA
                    + 8 TOKSFSE +  11 TPICHDE +  10 TPIDECH
                    + 12 TPIDESF  + 5 TPILASF +  10 TPISESF
                    + 5 TPISFLA +  8 TPISFSE  + 11 TWACHDE
                    + 18 TWACHLA +  18 TWASECH +  15 TWASELA
                    + 10 TWASESF +  5 TWASFLA  + 8 TWASFSE
     0    = SMOSE  =  SMOSE1 - TMOSECH - TMOSEDE - TMOSESF
     0    = SOKSE  =  SOKSE1 - TOKSESF
     0    = SPISF  =  SPISF2 - TPISFLA - TPISFSE
     0    = SWASE  =  SWASE1 - TWASECH - TWASELA - TWASESF
```

Figure 3-9. Re-entering SUBMAT To Change the Submatrix and List Its Equations

Sometimes we want to know how many rows and/or columns satisfy a conditional without necessarily displaying, listing or picturing them. This is done by the ADDRIM and COUNT commands, as shown in figure 3-10. We changed the submatrix with the conditionals given, but we now could examine it further, as with the PICTURE, LIST or DISPLAY commands. The row conditional specifies rows whose names begin with S and whose price is in $[.01,\infty)$, where the asterisk (*) denotes infinity in the value range `.01/*`.

```
ANALYZE ...C S P=.01/*,S
   Submatrix has  15 rows, 18 columns, and 18 nonzeroes
ANALYZE ...ADDRIM
 Submatrix has 18 COLS

          SUM          MIN          MAX        AVERAGE
      --------------------------------------------------
   L       0            0            0            0
   U     830.000        5.000      200.000       46.111
   C     553.000       10.000       60.000       30.722
   X     660.000        0          100.000       36.666
   D   - 125.000    -  23.000        4.000    -   6.944
      --------------------------------------------------
ANALYZE ...ADDR R
 Submatrix has 15 ROWS

          SUM          MIN          MAX        AVERAGE
      --------------------------------------------------
   L       0            0            0            0
   U       0            0            0            0
   C       0            0            0            0
   Y       0            0            0            0
   P     643.000       11.000       70.000       42.866
      --------------------------------------------------
```

Figure 3-10. The COUNT and ADDRIM Commands

Here is a breakdown of the first specification in figure 3-10.

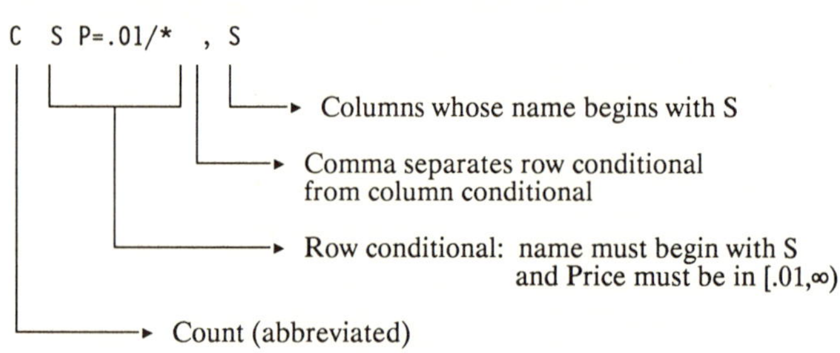

The COUNT gives the numbers of rows, columns and nonzeroes in the submatrix. In this case, the submatrix was set by the conditionals. The COUNT command can also be used without a conditional to get these numbers for the current submatrix, which may have been set by another command.

The ADDRIM gives the sum, minimum, maximum, and average for each rim value:

C = objective;
L = lower bound;
U = upper bound;
X = column level;
Y = row level (Y=AX);
P = row price;
D = column reduced cost (D=C-PA).

The operations skip infinite values; for example, the sum of the upper bounds shown applies only to the bounded (and fixed) columns in the submatrix.

The first ADDRIM gives the columns because that was the last choice in the DISPLAY (c.f., figure 3-5). To switch to the submatrix rows, the second ADDRIM specifies R. Alternatively, we could have specified

SWITCH DISP

(or simply SW D) to switch the display from column to row; then, we could follow this with ADDRIM to see the same result.

```
ANALYZE ... PI
                               Picture of submatrix
            S S S S S S S S S S S S S S S S S S
            M M O O O O O P P P P P P P P W W W
            O O K K K K K I I I I I I I I A A A
            S S C D L S S C C D D L S S S C S S
            E F H E A E F H H E E A E F F H E F
            1 1 1 1 1 1 1 1 2 1 2 1 1 1 2 1 1 1
SMOSE       +                                    =  0
SMOSF         +                                  =  0
SOKCH           +                                =  0
SOKDE             +                              =  0
SOKLA               +                            =  0
SOKSE                 +                          =  0
SOKSF                   +                        =  0
SPICH                     + +                    =  0
SPIDE                         + +                =  0
SPILA                             +              =  0
SPISE                               +            =  0
SPISF                                 + +        =  0
SWACH                                     +      =  0
SWASE                                       +    =  0
SWASF                                         +  =  0
```

Figure 3-11. Picturing the Submatrix Created by ADDRIMs

There is one more basic query command to see all information about activities: SHOW. As shown in figure 3-12, this shows solution, rim and matrix coefficient values. Before showing the columns, the SUB R * is used to put all rows in the submatrix. Otherwise, each coefficient table would have only the rows that were in the previous submatrix.

```
ANALYZE ... SUB R *
ANALYZE ... SHOW C TMO**DE D=0
 COL TMOSEDE  HAS 3 NONZEROES
 BOUNDS:            L=0            U=*
 SOLUTION:  STAT=B  X=20.000       D=0
 Row      Coefficient     Price       Factor
 ---------------------------------------------
 COST           20.000      - 1.0         20.000
 SMOSE         - 1.0         55.000       55.000
 DMODE           1.0         75.000    -  75.000
 =============================================
                              SUM:        0

 COL TMOSFDE  HAS 3 NONZEROES
 BOUNDS:            L=0            U=*
 SOLUTION:  STAT=B  X=0            D=0
 Row      Coefficient     Price       Factor
 ---------------------------------------------
 COST           12.000      - 1.0         12.000
 SMOSF         - 1.0         63.000       63.000
 DMODE           1.0         75.000    -  75.000
 =============================================
                              SUM:        0
```

Figure 3-12. Using SHOW COL to see Activity Information: Rim, Solution, Matrix Coefficients, and Reduced Cost Factors

The columns shown are those whose reduced costs are zero and whose name begins with TMO and ends with DE, leaving characters 4&5 free. The *factors* are the parts of the reduced cost: $-P_iA_{ij}$, where P_i is the price of the i-th row and A_{ij} is the matrix coefficient. By convention, $i=0$ corresponds to the objective row, which is COST in WOODNET. Because we are minimizing, $P_0=-1$, so its factor is $-(-1)c_j$, or simply c_j. A positive factor represents a cost of the activity, imputed from the optimal price times the associated coefficient, A_{ij}. A negative factor represents a revenue from the activity.

In a canonical minimization formulation (c.f., §2.1), the row constraint is of the form, $Ax \geq b$, which is equivalent to $L=b$ and $U=\infty$ for the row levels ($y=Ax$). Further, the optimal prices are non-negative, so a positive factor corresponds to a positive price times a positive coefficient. This agrees with the activity being a *producer* of the i-th domain for this instance. A negative factor corresponds to a positive price and a negative coefficient. This agrees with the activity being a *consumer* of the i-th domain for this instance. If either the price or the coefficient is zero, the activity is not producing or consuming anything of value.

Since all rows are shown, the sum of the factors is theoretically equal to the reduced cost (deviation caused by rounding and by formatted output). For example, column TMOSEDE has a COST factor of 20, which is just its input cost. The other 2 factors, 55 in row SMOSE and –75 in ROW DMODE, represent a cost and revenue, respectively. The sum is 0, which is precisely its reduced cost.

The form of output from the SHOW command can be varied. One option in showing columns is the ARRAY option, shown in figure 3-13.

```
ANALYZE ... SHOW //ARRAY
            TMOSEDE       TMOSFDE
            ------------  ------------
  LEVEL:        20.000         0
  LO:            0             0
  UP:            *             *
  COST:         20.000        12.000
  DJ:            0             0
  SMOSE        - 1.0
  DMODE          1.0           1.0
  SMOSF                      - 1.0
```

Figure 3-13. The ARRAY Option for SHOW COLUMN

There is a difference with the ARRAY option in the setting of the submatrix. In general, the SHOW command applies to a row *or* a column, so the submatrix is not reduced by removing null rows or columns, respectively. When the ARRAY option is specified, however, the submatrix is changed the same way as COUNT, LIST and PICTURE commands. In particular, if you specify SHOW COL T D=0, you will see all columns whose name begins with T and whose reduced cost is zero. If you put all rows in the submatrix first, you will see those having nonzero coefficients in these columns, and all rows remain in the submatrix after the columns are shown. With the ARRAY option, however, rows that do not have nonzero coefficients in these columns are removed from the submatrix. The output from the SHOW command will contain the same rows, but now the submatrix is changed, so DISPLAY ROW now displays only those rows that remained in the submatrix (i.e., that have a nonzero coefficient in at least one of the columns shown).

3.3 Syntax-based explanations.

Now that we have used basic query commands, we want to be able to translate results into English. This is done with a *syntax file*, prepared by the model builder. The structure of this file and how to create one for a model is postponed until chapter 6. Here we illustrate the commands that give syntax-based explanations.

Continuing with the same session, figure 3-14 demonstrates how to see an explanation of the rows and columns in the submatrix. After the specification, REA S, reads the syntax file into memory, the EXPLAIN command (abbreviated EXP) translates each column. The last rim query (c.f., figure 3-12) was for the column, and this is another global setting, like the conditional. To switch from column to row, we enter EXP R. The absence of a conditional for the rows preserves the submatrix.

Any set of rows and columns can be explained in this way. The EXPLAIN command has the same conditional specification as the DISPLAY and ADDRIM commands. (Of course, if we add a conditional, the submatrix changes.) We next see the overall syntax on which the explanation is based.

```
ANALYZE ... PIC
                                  Picture of submatrix
         T T
         M M
         O O
         S S
         E F
         D D
         E E
COST     + + = MIN
DMODE    + + =  0
SMOSE    -   =  0
SMOSF      - =  0
ANALYZE ... REA S
 Syntax file read in
ANALYZE ... EXP
   Column TMOSEDE transports mahogany from Seattle to Denver.
   Column TMOSFDE transports mahogany from San Francisco to Denver.
ANALYZE ... EXP R
   Row COST is the objective (MINIMIZE).
   Row DMODE balances demand of mahogany in Denver.
   Row SMOSE balances supply of mahogany in Seattle.
   Row SMOSF balances supply of mahogany in San Francisco.
```

Figure 3-14. Explaining Rows and Columns

Figure 3-15 illustrates the SYNTAX command with the * to indicate that we want to see the sets, rows and columns. This is generally a useful way to become familiar with someone else's model. Whereas the SUMMARY command gives LP statistics, usually meaningful to an LP expert, the SYNTAX command gives a summary of the model's structure. (In the next section we illustrate a schema view that gives another way to use model syntax for gaining insight into the overall model structure.)

```
ANALYZE ... SYN *
  3 Sets...
   Set MT = material
            PI = pine
            WA = walnut
            OK = oak
            MO = mahogany
   Set SR = location
            DE = Denver
            CH = Chicago
            SF = San Francisco
            LA = Los Angeles
            SE = Seattle
   Set DR = location
            DE = Denver
            CH = Chicago
            SF = San Francisco
            LA = Los Angeles
            SE = Seattle

Row syntax has 2 classes
  A row that begins with S balances supply of some material in some
      location.
  A row that begins with D balances demand of some material in some
      location.

Column syntax has 3 classes
  A column that begins with S supplies some material at some location.
  A column that begins with T transports some material from some
      location to some location.
  A column that begins with D demands some material at some location.
```

Figure 3-15. Displaying the Model Row and Column Syntax

We see that rows and columns are defined by a syntax of the form *c*(*domain*), where *c* is a class descriptor (like S or D) and the *domain* is a cross product of sets (often restricted, such as in a distribution network where not every supplier can send to every consumer). Sometimes it is desirable to extract classes from a domain. This is the function of the TALLY command, as shown in figure 3-16. After setting the submatrix to the entire set of rows and columns with the SUB * specification, TAL *?? asks to tally the names for the class associated with characters 2&3 (members of set MT in this case).

```
ANALYZE ... SUB *
ANALYZE ... TAL *??
   36 ROWS TALLIED
    5 ROW CLASSES IN SUBMATRIX
   1 in *OS          7 in *MO          10 in *OK          10 in *PI
   8 in *WA
```

Figure 3-16. Tallying Classes

The rows, rather than the columns, were tallied because that is what we last did with the EXP R (c.f., figure 3-14). If we want to tally columns, we specify TAL *?? C. The first specification after the tally command (abbreviated TAL) is the *query mask*. This is a name mask with the question mark (?) used as a query character - that is, the field in the name we want to tally. Then, the ROW or COLUMN and conditional follow the same form as in the DISPLAY, ADDRIM, SHOW and EXPLAIN commands. Since we did not specify a conditional, the submatrix remains equal to the entire matrix. The first of the 5 classes found is really not a class; it is the COST row, which is why it has only 1 member in the class *OS. The other 4 classes are the materials (MT), as shown in figure 3-15.

One thing we can do with this information is trace a flow of materials. From the SYNTAX and TALLY information, we know mahogany (MO) is a member of the set of materials (MT). We can trace its flow by putting into the submatrix all pertinent rows and all columns that have positive levels (that *intersect* at least one of the rows). This appears in figure 3-17 with the PICTURE command (abbreviated `PIC`). The first name mask, `*MO`, delineates rows that have `MO` in character positions 2&3. Following the comma, the column conditional begins with the name mask * to specify all names. Then, we restrict the level of the columns to be pictured by having a level of at least 1; this is specified by the rim value range, `X=1/*` (in the WOODNET model all activity levels are integer-valued, so this specification is equivalent to a positive level; more generally, one may choose a small positive value for the range, like .00001/*). In addition to picturing, we have also set the resulting submatrix.

```
ANALYZE... PIC *MO,* X = 1/*
                                 Picture of submatrix
          D D D D D S S T T T T T
          M M M M M M M M M M M M
          0 0 0 0 0 0 0 0 0 0 0 0
          C D L S S S S S S S S S
          H E A E F E F E E E F F
          1 1 1 1 1 1 1 C D S L S
                        H E F A E
 DMOCH    -             +           = 0
 DMODE      -             +         = 0
 DMOLA        -               +     = 0
 DMOSE          -               +   = 0
 DMOSF            -         +       = 0
 SMOSE              +   - - -       = 0
 SMOSF                +       - -   = 0
```

Figure 3-17. A Flow of Materials (MO)

Figure 3-18 then proceeds to add the COST row to the submatrix, but this time we avoid entering the SUBMAT command level. Instead, we use the direct form of SUBMAT. The specification `A R C` abbreviates `ADD ROW COST` (the COST row is the only one that begins with C). The picture shown does not give the numerical values of the activity costs, so we want to LIST the COST row over the submatrix columns.

By specifying the COST row in the LIST command, we have reduced the submatrix to contain only this row, and we lost the columns whose costs are 0 – that is, those of the Demand activities. (Specify PIC yourself to see what the submatrix is now.)

```
ANALYZE ... SUBMAT A R C
ANALYZE ... PI

                                Picture of submatrix
         D D D D D S S T T T T T
         M M M M M M M M M M M M
         0 0 0 0 0 0 0 0 0 0 0 0
         C D L S S S S S S S S S
         H E A E F E F E E E F F
         1 1 1 1 1 1 1 C D S L S
                       H E F A E
 COST              + + + + + + + = MIN
 DMOCH   -               +         =   0
 DMODE     -               +       =   0
 DMOLA       -                 +   =   0
 DMOSE         -                 + =   0
 DMOSF           -           +     =   0
 SMOSE             +   - - -       =   0
 SMOSF               +         - - =   0
ANALYZE... L COST
     MIN    COST    =   55 SMOSE1 +  45 SMOSF1  + 18 TMOSECH
                    +   20 TMOSEDE +  10 TMOSESF +  5 TMOSFLA
                    +   8 TMOSFSE
```

Figure 3-18. Picture of Flow with COST Row Added to Submatrix

Syntax-based explanations of rows and columns can be included with other commands. Enter SWITCH to see the switches, and note the one called SYNTAX, which is off (value=F). To illustrate, figure 3-19 uses the DISPLAY command with the SYNTAX switch on. Note that the syntax can be suppressed, regardless of the switch setting, by specifying //NOSYNTAX as an option.

```
ANALYZE ...SWITCH SYNTAX
ANALYZE ...DISPLAY ROW DMO
  5 Rows are in submatrix.
  ROW       STAT  LEVEL (Y)     LO_BOUND      UP_BOUND      PRICE (P)
  ------------------------------------------------------------------
  DMOCH      L         0            0             0            73.000
   ...balances demand of mahogany in Chicago
  DMODE      L         0            0             0            75.000
   ...balances demand of mahogany in Denver
  DMOLA      L         0            0             0            68.000
   ...balances demand of mahogany in Los Angeles
  DMOSE      L         0            0             0            71.000
   ...balances demand of mahogany in Seattle
  DMOSF      L         0            0             0            65.000
   ...balances demand of mahogany in San Francisco
ANALYZE ...DISP  //NOSYNTAX
  5 Rows are in submatrix.
  ROW       STAT  LEVEL (Y)     LO_BOUND      UP_BOUND      PRICE (P)
  ------------------------------------------------------------------
  DMOCH      L         0            0             0            73.000
  DMODE      L         0            0             0            75.000
  DMOLA      L         0            0             0            68.000
  DMOSE      L         0            0             0            71.000
  DMOSF      L         0            0             0            65.000
```

Figure 3-19. Displaying Syntax with Query

We can also see syntax if the SYNTAX switch is off, as illustrated in figure 3-20. We first turn the SYNTAX switch off, then specify PRICE and SYNTAX as options for the display. Note that now the syntax-based explanation is given on the same line as each row, rather than on a separate line as in figure 3-19. ANALYZE uses the screen width settıng to determine if the explanation can fit on the same line. If it can fit, it is put on the same line; otherwise, it is put on a separate line.

```
ANALYZE ...SWITCH SYNTAX
ANALYZE ...D  //PRICE,SYNTAX
  5 Rows are in submatrix.
  ROW       PRICE (P)
  ---------------------
  DMOCH         73.000 balances demand of mahogany in Chicago
  DMODE         75.000 balances demand of mahogany in Denver
  DMOLA         68.000 balances demand of mahogany in Los Angeles
  DMOSE         71.000 balances demand of mahogany in Seattle
  DMOSF         65.000 balances demand of mahogany in San Francisco
```

Figure 3-20. Displaying Syntax without SYNTAX Switch

The SYNTAX option can be used with the SHOW command as well. Figure 3-21 gives an example, using a general mask to specify all activities that begin with TMO (the fixed mask portion), and have SE somewhere after position 3 (the sliding mask portion, indicated by the "). This specifies activies for transporting mahogany into or out of the Southeast. The conditional further specifies only those activities with levels in the 10 to 20 (inclusive).

```
ANALYZE ...SUB *
ANALYZE ...SHO COL  TMO"SE  X=10/20
  COL TMOSEDE  HAS 3 NONZEROES...transports mahogany from Seattle to Denver
  BOUNDS:               L=0            U=*
  SOLUTION:  STAT=B  X=20.000        D=0
  Row      Coefficient     Price       Factor
  -----------------------------------------------
  COST           20.000       - 1.0         20.000 is the objective (MINIMIZE)
  SMOSE          - 1.0        55.000        55.000
  ...balances supply of mahogany in Seattle
  DMODE            1.0        75.000      - 75.000
  ...balances demand of mahogany in Denver
  ===============================================
                                SUM:         0

  COL TMOSFSE  HAS 3 NONZEROES...transports mahogany from San Francisco to
  Seattle
  BOUNDS:               L=0            U=*
  SOLUTION:  STAT=B  X=10.000        D=0
  Row      Coefficient     Price       Factor
  -----------------------------------------------
  COST            8.000       - 1.0          8.000 is the objective (MINIMIZE)
  SMOSF          - 1.0        63.000        63.000
  ...balances supply of mahogany in San Francisco
  DMOSE            1.0        71.000      - 71.000
  ...balances demand of mahogany in Seattle
  ===============================================
                                SUM:         0
```

Figure 3-21. Using SYNTAX with the SHOW Command

3.4 Schema view.

The syntax may also be used to get a schema view, as shown in figure 3-22. The table's rows are called *row strips*, and the columns are called *column strips*. Each strip corresponds to a set of rows or columns, distinguished by the first letter of its name. Then, the *domain* of that strip is the product of sets over which it is defined. The cell entries are the ranges of values in this instance of the model. Typically, one finds ±1's as part of the model structure, independent of the instance. A blank cell means there are no nonzero coefficients in that submatrix. The range of values in the COST of activities T(MT,SR,DR) is shown as `5/20`, which means each nonzero is in the interval [5, 20].

```
ANALYZE ... SCHEMA DISPLAY TABLE,DOMAINS
           :::::::::::::::: BLOCK SCHEMA ::::::::::::::::
             S(MT,SR)     T(MT,SR,DR) D(MT,DR)     HOMOGEN
  S(MT,SR)  1             -1                         = 0
  D(MT,DR)                1           -1             = 0
  COST       10/60        5/20                     ...MIN
  :LO        0            0           5/100
  :UP        5/*          *           =

         DOMAIN INFORMATION
   MT        material
   SR        location
   DR        location
```

Figure 3-22. A Schema View

The lower and upper bounds are also shown as the last 2 strips. All Supply activities (i.e., whose column name begins with S) have lower bounds of 0 and upper bounds in the interval $[5, \infty)$. All Demand activities (i.e., whose column name begins with D) are fixed in the interval [5, 100], where the = in the :UP strip indicates that each upper bound equals its lower bound. All Transportation activities (i.e., whose column name begins with T) are simply non-negative.

In general, a schema gives another view of a linear program, where one can see how equations and activities are related to each other and the domains over which they are defined. The special row strips, :LO and :UP, give lower and upper bounds of each activity class, represented by column strips. The other row strips are equation classes, and a cell entry is a term in the equation. For the schema in figure 3-22, the linear programming model is:

Minimize COST = $\Sigma_{mi}\, c_{mi} S_{mi} + \Sigma_{mij}\, t_{mij} T_{mij}$ subject to:

$S_{mi} - \Sigma_j T_{mij} = 0$ for $m \in MT$, $i \in SR$ (Supply equation, S(m, i))
$D_{mj} - \Sigma_i T_{mij} = 0$ for $m \in MT$, $j \in DR$ (Demand equation, D(m, j))
$0 \le S \le U$, $0 \le T$, $D = d$ (Bounds).

According to the schema, the data for the linear program are the costs and bounds. Domains are products of the sets of materials (MT), supply regions (SR) and demand regions (DR). The appearance of 1's in the schema are logical coefficients.

In this chapter we have gone through the basic query commands. One may group these by the type of conditional:

Row or Column:	`ADDRIM, DISPLAY, EXPLAIN, SHOW, TALLY`
Row and Column:	`COUNT, LIST, PICTURE`

We also illustrated how to obtain overviews of the model with the SUMMARY, SYNTAX and SCHEMA commands; and, we showed how to use the SUBMAT command level when a simple conditional is not enough to obtain the submatrix of interest. Besides the HELP command, see CONDNL.DOC, which gives on-line help for conditional specifications.

Some of the session control commands were used (such as SCREEN), and we used the READIN command to read WOODNET into memory. At this point, you might print QUICKREF.DOC (which is in ANALYZE.DOC if you EXECUTEd DOCANAL, as suggested). This will give an overview of these and other commands and the functions they serve.

In the next chapter we continue to use these commands for specific analysis problems. This will serve to reinforce what we have covered, and it will introduce additional capabilities for computer-assisted analysis. If you leave ANALYZE and want to proceed with the exercises later, simply enter QUIT (or Q). Later, enter EXEC WOODNET at the ANALYZE prompt. This reads the LP into memory (matrix + solution), then it reads the syntax file, which we shall use in explaining the analytical results in English.

Chapter 4. Examples of Analysis

Detection is, or ought to be, an exact science and should be treated in the same cold and unemotional manner.

Sherlock Holmes

In this chapter we show how to use ANALYZE for some exercises. The first is a price interpretation, continuing with WOODNET where we left off in the previous chapter. After solving the analysis problem algebraically - that is, as LP experts - we use the model's syntax to put the final explanation in English. We then demonstrate how this can be automated with a *rule file* (see §6.7) to support the INTERPRT command. After analyzing an optimal instance of WOODNET, we diagnose an infeasibility in another instance of the WOODNET model, and the last exercise is to diagnose an unbounded instance.

4.1 Price interpretation for the LP expert.

We begin with an exercise to explain the $73 *Price* for the first demand row shown in figure 3-4, namely DMOCH. Before we do so, enter EXEC WOODNET at the ANALYZE prompt. Your screen should appear as in figure 4-1.

```
ANALYZE ...EXEC WOODNET
 WOODNET.....an optimal network model
   Analyze the properties,  beginning with model attributes.
   Then,  try using the  TRACE and RATEOF commands.  Try  to
   explain what you learn about the model, using the EXPLAIN
   and SYNTAX commands to put your  explanation  in English.
   In particular,  explain the marginal prices of the supply
   and demand rows.
```

Figure 4-1. Executing WOODNET (Reads in LP Packed and Syntax files)

Referring back to figure 3-4, we want to explain the price of $73 for the first demand row (DMOCH). We begin, as shown in figure 4-2, by picturing the submatrix composed of this row and all basic columns. The PICTURE command removes null rows and columns from the submatrix before giving the picture, so we see the resulting 1×1 submatrix. Among the basic activities, only activity TMOSECH has a nonzero in row DMOCH.

At this point, we may expand the submatrix, perhaps with the SUBMAT command, to bring other rows that intersect the one column. This, in turn, may lead to bringing in other columns, preferably basic, until we eventually reach a *complete* submatrix. This has been automated with the TRACE command. Starting with the

1×1 submatrix, TRACE seeks to balance the equations with basic activities (see §9.2). When it is finished, a submatrix is created that contains the original one and identifies the portion of the LP relevant to the flows that determine the marginal price.

```
ANALYZE ... PICTURE DMOCH,* S=B
                                Picture of submatrix
          T
          M
          O
          S
          E
          C
          H
DMOCH     + =  0
ANALYZE ... TRACE
 TRACE complete...submatrix is set
```

Figure 4-2. Beginning an Analysis to Explain the $73 Price of Row DMOCH

Then, we picture the resulting submatrix, shown in figure 4-3. In this case, the 3×3 submatrix shows the COST and a flow from the supply activity SMOSE1, shipped by the transportation activity TMOSECH, and finally consumed by the demand activity, DMOCH1.

```
ANALYZE ...PI
                                Picture of submatrix
          D S T
          M M M
          O O O
          C S S
          H E E
          1 1 C
              H
COST        + + = MIN
DMOCH     -   + =  0
SMOSE       + - =  0
```

Figure 4-3. Picturing the Submatrix Found by TRACE

The picture is useful for seeing patterns and having a quick look at a submatrix, but we need the numerical values of the nonzeroes to proceed. This is done with the LIST command, as shown in figure 4-4.

The first LIST command specification results in an equation listing, which can be regarded as an algebraic view, and it is the default form of output. The second listing of the same submatrix specifies //ROW to obtain a tabular list of the nonzeroes, sorted by rows. The third listing specifies //COL to obtain a tabular list, sorted by columns.

```
ANALYZE ...L
      MIN    COST    =   55 SMOSE1 +  18 TMOSECH
      0     = DMOCH  =- DMOCH1 + TMOSECH
      0     = SMOSE  =  SMOSE1 - TMOSECH
ANALYZE ...L //ROW
  COST      SMOSE1          55.000
            TMOSECH         18.000
  DMOCH     DMOCH1         - 1.0
            TMOSECH          1.0
  SMOSE     SMOSE1           1.0
            TMOSECH        - 1.0
ANALYZE ...L //COL
  DMOCH1    DMOCH          - 1.0
  SMOSE1    SMOSE            1.0
            COST            55.000
  TMOSECH   COST            18.000
            SMOSE          - 1.0
            DMOCH            1.0
```

Figure 4-4. Listing the Nonzeroes in the Submatrix (Three Forms)

Figure 4-5 shows the result of a display command to see the 3 columns in the submatrix, primarily their levels. The demand activity (namely, DMOCH1) is at its fixed level. The associated transportation activity (namely, TMOSECH) is thus at this level (25), but the supply activity (SMOSE1) level is 75 because it supplies other demands.

```
ANALYZE ... D C
 3 Columns are in submatrix.
 COL      STAT  LEVEL (X)     LO_BOUND      UP_BOUND      PRICE (D)
 ------------------------------------------------------------------
 DMOCH1   L        25.000       25.000        25.000         73.000
 SMOSE1   B        75.000        0              *              0
 TMOSECH  B        25.000        0              *              0
```

Figure 4-5. Displaying the Columns

You might now have the image of figure 4-6 in your mind. We know from the algebraic substructure - that is, the submatrix found - that there is a *margin-setting* path from supply in Seattle (SE) to demand in Chicago (CH). The supply and demand boxes may be regarded as the nodes associated with the balance rows (namely, SMOSE and DMOCH). The arrows represent the 3 activities. The first activity (SMOSE1) supplies mahogany (MO) in Seattle. Its input cost of $55 and its (optimal) level of 75 units are shown above the first arc. The second activity (TMOSECH) transports 25 units of the mahogany from Seattle to Chicago at its unit input COST of $18. The third activity (DMOCH1) consumes the 25 units of mahogany at zero COST. The imputed consumer's cost, however, is $73 per unit, shown below its arc. The row prices, P=55 for the Seattle supply balance and P=73 for the Chicago demand balance, are shown below their boxes.

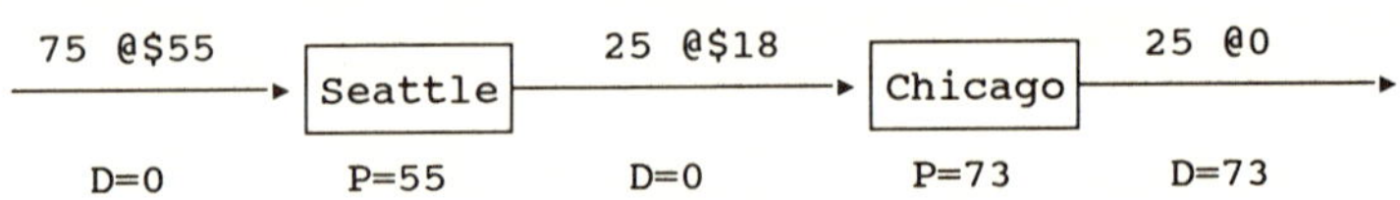

Figure 4-6. A Flow Diagram Corresponding to the Submatrix

We are now in a position to give an algebraic answer to this exercise. The TRACE reveals a path from supply to demand; the levels of the activities in this path may be perturbed to accommodate a change in the right-hand side of the demand row DMOCH (equivalently, the fixed level of the associated demand activity, DMOCH1). The $73 is the total unit cost of this perturbation. The last listing (c.f., figure 4-4) shows that the total (input) cost is $55 for the supply activity (SMOSE1) and $18 for the transportation activity (TMOSECH). Because there are no loss or gain factors (i.e., the body coefficients are ±1), the actual delivered cost is $55 + $18, or $73.

It is not usually the case that the delivered cost equals the marginal price. We next analyze such a case by performing the same exercise for row DMOLA. The problem is to explain the $68 in terms of the model/solution structure and the data. Figure 4-7 starts with the same PICTURE and TRACE specifications for this row as we did in figure 4-2 for the first row.

```
ANALYZE ... PI DMOLA,* S=B
                             Picture of submatrix
         T
         M
         O
         S
         F
         L
         A
DMOLA    + =  0
ANALYZE ... TRACE
 TRACE complete...submatrix is set
ANALYZE ... PI
                             Picture of submatrix
         D S T
         M M M
         O O O
         L S S
         A F F
         1 1 L
             A
COST       + + = MIN
DMOLA    -   + =  0
SMOSF      + - =  0
```

Figure 4-7. Beginning the Price Trace for Row DMOLA

Now we DISPLAY the columns, as shown in figure 4-8. We see that the supply activity (SMOSF1) is at its bound (S=U), which is 30. Upon listing the COST row (using EQUATION format), we see that its delivered cost is only $50. The difference of $50 and $68 is what we seek to explain. This is precisely the reduced cost of the capacitated supply activity, which is displayed as −18.

Before proceeding, notice that the last LIST specification caused the submatrix to drop the other equations. Only the COST row remains in the submatrix, because we specified LIST COST, which entered a row conditional: rows whose name begins with COST. This is why only the COST equation was listed, and with only those columns that were already in the submatrix, since no new column conditional was specified.

```
ANALYZE ... D C
 3 Columns are in submatrix.
 COL      STAT  LEVEL (X)    LO_BOUND     UP_BOUND     PRICE (D)
 ---------------------------------------------------------------
 DMOLA1   L         20.000       20.000       20.000        68.000
 SMOSF1   U         30.000       0            30.000      - 18.000
 TMOSFLA  B         20.000       0            *             0
ANALYZE ... D R
 3 Rows are in submatrix.
 ROW      STAT  LEVEL (Y)    LO_BOUND     UP_BOUND     PRICE (P)
 ---------------------------------------------------------------
 COST     B     25135.000      - *          *            - 1.0
 DMOLA    L         0            0            0             68.000
 SMOSF    L         0            0            0             63.000
ANALYZE ... L COST
     MIN    COST    =  45 SMOSF1 +  5 TMOSFLA
```

Figure 4-8. Display of Columns and Rows in Path and List of COST Row (Delivered Cost = 50)

We must now trace the flow out of this supplier to explain its reduced cost. There is a variety of ways to proceed. We begin to explain the $18 difference by seeing who else receives the other 10 units of supply from San Francisco. Figure 4-9 shows the result of picturing the associated supply row (SMOSF) with the columns of not only basic activities, but also of all columns with zero reduced cost. (This distinction will become apparent, as our problem needs to include the dual degeneracy when tracing prices.) As shown at the top of figure 4-9, there are 3 columns whose associated activities have D=0 and that intersect row SMOSF. We then TRACE this submatrix, but before picturing the result, we delete the COST row from the submatrix. Our purpose in doing so at this point is merely to illustrate the direct use of the SUBMAT command. In other cases we may want to delete all free rows to reduce the size of the submatrix before picturing it. This is done by specifying:

```
SUBMAT DELETE ROWS * L=-* U=*
```

or simply

```
SUB D R * L=-* U=*
```

```
ANALYZE ... PI SMOSF,* D=0
                             Picture of submatrix
        T T T
        M M M
        0 0 0
        S S S
        F F F
        D L S
        E A E
SMOSF   - - - =  0
ANALYZE ... TR
 TRACE complete...submatrix is set
ANALYZE ... SUB D R COST
ANALYZE ... PI
                             Picture of submatrix
        D D D S T T T
        M M M M M M M
        0 0 0 0 0 0 0
        D L S S S S S
        E A E F F F F
        1 1 1 1 D L S
                E A E
DMODE   -       +     =  0
DMOLA     -       +   =  0
DMOSE       -       + =  0
SMOSF         + - - - =  0
```

Figure 4-9. Trace of Margin-Setting Flows from the San Francisco Supplier of Mahogany

We want to see the flows and the reduced costs on the activities in this submatrix. To do so, we DISPLAY the rows and columns, shown in figure 4-10. Our story starts to take shape, as the San Francisco supplier ships to other places. Since we know the supply bound (U=30) is binding, the $18 must be due to the value of the capacity in fulfilling some other demand(s). That is the story we want to develop.

```
ANALYZE ... D R
 4 Rows are in submatrix.
 ROW      STAT  LEVEL (Y)     LO_BOUND      UP_BOUND      PRICE (P)
 -----------------------------------------------------------------
 DMODE     L        0             0             0            75.000
 DMOLA     L        0             0             0            68.000
 DMOSE     L        0             0             0            71.000
 SMOSF     L        0             0             0            63.000
ANALYZE ... D C
 7 Columns are in submatrix.
 COL      STAT  LEVEL (X)     LO_BOUND      UP_BOUND      PRICE (D)
 -----------------------------------------------------------------
 DMODE1    L        20.000        20.000        20.000       75.000
 DMOLA1    L        20.000        20.000        20.000       68.000
 DMOSE1    L        10.000        10.000        10.000       71.000
 SMOSF1    U        30.000        0             30.000     - 18.000
 TMOSFDE   B        0             0             *            0
 TMOSFLA   B        20.000        0             *            0
 TMOSFSE   B        10.000        0             *            0
```

Figure 4-10. Displaying Rows and Columns in Submatrix

At this point, it helps to know what the rows and columns mean. Figure 4-11 gives their syntax with the EXPLAIN command. We purposely did not LIST COST right after the displays in figure 4-10 because to do so changes the submatrix, and we first want to explain its rows and columns.

```
ANALYZE ... EXP R
    Row DMODE balances demand of mahogany in Denver.
    Row DMOLA balances demand of mahogany in Los Angeles.
    Row DMOSE balances demand of mahogany in Seattle.
    Row SMOSF balances supply of mahogany in San Francisco.
ANALYZE ... EXP C
    Column DMODE1 demands  mahogany at Denver.
    Column DMOLA1 demands  mahogany at Los Angeles.
    Column DMOSE1 demands  mahogany at Seattle.
    Column SMOSF1 supplies mahogany at San Francisco.
    Column TMOSFDE transports mahogany from San Francisco to Denver.
    Column TMOSFLA transports mahogany from San Francisco to Los Angeles.
    Column TMOSFSE transports mahogany from San Francisco to Seattle.
```

Figure 4-11. Explaining the Syntax of the Rows and Columns

Now we want to see the costs, which we list (with row tabular form) in figure 4-12. This tells us the delivered costs to Denver, LA and Seattle. That is, we already know the supply cost in San Francisco is $45. Now the TRACE and the earlier EXPLAINs tell us that there are 3 activities with zero reduced cost. The first is TMOSFDE, which transports mahogany from San Francisco to Denver at a unit cost of $12. The second is TMOSFLA, which transports mahogany from San Francisco to Los Angeles at a unit cost of $5, which is the one we already saw when we started this exercise. The third is TMOSFSE, which transports mahogany from San Francisco to Seattle at a unit cost of $8.

```
ANALYZE ... L COST //R
 COST     SMOSF1        45.000
          TMOSFDE       12.000
          TMOSFLA        5.000
          TMOSFSE        8.000
```

Figure 4-12. Seeing the Costs of the Activities

From figure 4-10, we see that only two of these margin-setting flows – that is, where D=0 – are positive. The Status of activity TMOSFDE is Basic, but its level is 0. We could explore the link with Seattle, but it is this degeneracy that holds the key for the answer we seek. Figure 4-13 pictures the submatrix with the associated demand row (DMODE) and all columns whose activity level is at least 1 (indicated by X=1/*). We obtain the answer to the question, "Where does Denver gets its mahogany?"

```
ANALYZE ... PI DMODE,* X=1/*
                        Picture of submatrix
         D T
         M M
         O O
         D S
         E E
         1 D
           E
DMODE    - + =  0
ANALYZE ... TR
 TRACE complete...submatrix is set
ANALYZE ... L COST
     MIN    COST    =    55 SMOSE1 +  20 TMOSEDE
```

Figure 4-13. Tracing the Price of Mahogany in Denver
(Price of Row DMODE)

Skipping the new picture after the trace, we immediately list the costs and find two things. First, the mahogany comes from only one supplier, namely Seattle. Second, the delivered cost is $75 ($55 supply, which is activity SMOSE1, and $20 transportation, which is TMOSEDE). You can DISPLAY the columns now to learn that this is like the first case, where Denver receives all of its mahogany from a single supplier, and there are no binding capacity constraints in the flow path, so the marginal price of row DMODE is its delivered cost of $75.

Now we can put the pieces together and construct the analysis story. Figure 4-14 shows a flow graph that describes the situation. The San Francisco supplier sends to Los Angeles (where we started) and to Seattle. The latter flow is irrelevant to our original exercise of explaining the $68 marginal price for row DMOLA because it is not the cause of the $18 economic rent, which shows up as the –18 reduced cost of the supply activity (SMOSF1). This is due to the margin-setting property in its link with Denver, even though there is no positive flow across that arc (activity TMOSFDE).

Advice from a student

Economic terms, like "economic rent" and "margin-setting", are beyond the present scope. If you are just learning linear programming, this is a good time to think about what the terms mean and about how we have conducted analyses.

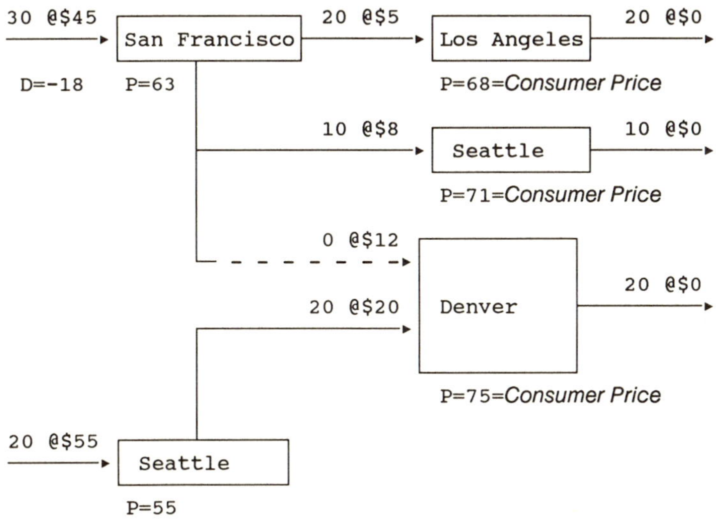

Figure 4-14. A Graph of the Final Analysis

The story is that the demand price for mahogany in Denver, which is P=75 for row DMODE, is determined as in the first case: it receives its shipment from another supplier, namely from Seattle (via the transportation activity TMOSEDE), with delivered cost of $75. The presence of the link from SF, however, allows displacement if the capacity of the LA supplier would increase. The (actual) cost of shipping from San Francisco to Denver is $57 ($45 supply + $12 transportation), so the savings from displacement would be $18. This is what determined the economic rent we have sought to explain.

There are a number of ways to compose the final explanation. One is that the consumer's price of mahogany in LA ($68) is due to demand in Denver and the limited supply in San Francisco. Even if we reduce the production cost and the transportation cost from San Francisco to LA, the price would still be $68 due to the pressure resulting from limited supply capacity. The only way to reduce the $68 price is to reduce the delivered cost from Seattle to Denver!

4.2 Automatic interpretation.

The analysis we just did can be automated to a great extent, at least for the WOODNET model. First, let us return to the simpler case by putting the row DMOCH into the submatrix, as shown in figure 4-15, using the SUBMAT command (without leaving the main command level). We use the INTERPRT command, which requires a *rule file* (see §6.7). Note that it gives the same answer we obtained with the ANALYZE commands.

```
ANALYZE ... SUB R DMOCH
ANALYZE ... INTERP PRICE
  Row DMOCH balances demand of mahogany in Chicago.
 I shall try to interpret its price of 73.  The consumers in Chicago
receive  all of their mahogany from Seattle with delivered cost = 73.
Thus, marginal price = delivered cost = 73 (from Seattle)  = price of
DMOCH = Supply cost (55)  + Transportation cost (18).
```

Figure 4-15. Automatic Price Interpretation When Price = Delivered Cost

A Discrepancy

> In figures 4-15 and 4-16, the output display differs from your screen to accommodate the width of the User's Guide page, which is less than the width of your screen (unless you reduce the width with the SCREEN command).

The second example is more complex, and figure 4-16 shows that the demand price interpretation is not complete. The analyst, however, is advised to interpret the supply price because once the $18 *economic rent* is explained, the rest follows. (The term *economic rent* is used, and this may not be understood by the analyst. It depends on how the model builder (or manager) wrote the rule file.) As advised, we proceed to interpret the supply price with another use of the INTERPRT command (abbreviated in each of the two specifications). Note that we do not have to use the SUBMAT command. Instead, we name the row after specifying PRICE as what we want to interpret. The result is a successful interpretation, as we did before (but now automatically when we're not there for someone else with the same question).

```
ANALYZE ... INTERP PRICE DMOLA
  Row DMOLA balances demand of mahogany in Los Angeles.
 I shall try to interpret its price of 68.  The consumers in Los
Angeles receive all of their mahogany from San Francisco with
delivered cost = 50 (this supplier, however, has an economic rent of
18).  The marginal price (68)  = delivered cost + rent.  You might
try interpreting the supply price. .. see row SMOSF.
ANALYZE ... I P SMOSF
  Row SMOSF balances supply of mahogany in San Francisco.
 I shall try to interpret its price of 63.  The (input) supply cost =
45 (excluding economic rent = 18).  This supply is delivered to 2
consumers -- namely, to Los Angeles  with transportation cost = 5;
and, to Seattle with transportation cost = 8.
This supplier does not send any mahogany to Denver,  but it is this
consumer that is setting the supplier's economic rent.  That is, any
increase in the supply of mahogany in San Francisco would go to
Denver and displace its most expensive current delivered cost.
```

Figure 4-16. Automatic Price Interpretation When Price ≠ Delivered Cost

The complete interpretation may be done with chaining. For this case, we could automatically invoke the interpretation of the San Francisco supply price after the conclusion of the LA demand price. Presently, this is risky for general cases because there is no safeguard against cycling. Unlike the present case, there could be a cyclic structure that cannot be simply decomposed into a causal chain (see §6.7 for an explanation of how construct your own *rule file* to enrich and adapt the INTERPRT command).

Warning

> We have used the TRACE command successfully in the WOODNET model (both directly and from rule files), but the procedure is a heuristic that may not give meaningful results in other types of formulations. The particular property of WOODNET is that the equations are homogeneous, and activities are used to represent supply, demand and transportation activities. It is the qualitative property of the sign pattern resulting from this formulation that made the TRACE procedure work as it was designed (see §9.2 for a full discussion of this procedure).

4.3 A case of infeasibility.

Now we analyze an infeasible instance of WOODNET. To begin, we read in the packed LP file plus the syntax file with an execution file that was prepared for this exercise. This is shown in figure 4-17. (Whenever a new LP is read in, it replaces the old one. The submatrix is cleared, but session controls, such as screen settings, remain unchanged.)

```
ANALYZE ... EXEC WOODINFE
   WOODINFE.....an infeasible network problem.
   Try to diagnose the cause of infeasibility by first
   isolating a relevant portion of the model instance.
   Ultimately, you want to answer, "Why is this infeasible?"
   to someone who does not know LP.  Use the EXPLAIN command
   to put your explanation into English.  As a last resort use
   the REDUCE command, which should reveal the problem.
```

Figure 4-17. Loading an Infeasible Case

A reasonable place to start is to see what rows are infeasible. This is done at the top of figure 4-18 with the DISPLAY command. The specification uses D to abbreviate DISPLAY and R to abbreviate ROW. The * indicates that any name matches, and S=I means solution status is infeasible (level violates a bound). The response, we see, is that there are no infeasible rows in this solution. So, we must have some infeasible columns, and we display them.

```
ANALYZE ... D R * S=I
 0 Rows are in submatrix.
ANALYZE ... D C * S=I
 2 Columns are in submatrix.
 COL      STAT  LEVEL (X)    LO_BOUND     UP_BOUND     PRICE (D)
 ---------------------------------------------------------------
 DMOSE1    I         0         10.000         *            1.0
 DWASE1    I         0          5.000         *            1.0
ANALYZE ... PIC *
                            Picture of submatrix
         D D
         M W
         O A
         S S
         E E
         1 1
DMOSE    -   =  0
DWASE      - =  0
ANALYZE ... SUB COL *
ANALYZE ... PIC
                            Picture of submatrix
         D D
         M W
         O A
         S S
         E E
         1 1
DMOSE    -   =  0
DWASE      - =  0
```

Figure 4-18. Analysis of Infeasibility

The two infeasible columns displayed are demands of 10 and 5, respectively. The model should contain a path from supply to satisfy these demands, so we next picture the submatrix of these columns and all rows. This is done with the `PIC *` specification. Because no column conditional is specified, the submatrix columns are unchanged; the * for the row conditional, however, puts in all rows having a nonzero coefficient in one of these columns.

Then, we specify SUB COL * to enter into the submatrix all columns having nonzero coefficients in one (or both) of these rows. The subsequent picture is the same because no other columns intersect these rows.

The explanation is at hand: the demand equations (DMOSE and DWASE) force the associated demand activity levels (DMOSE1 and DWASE1, respectively) to be 0, probably due to inadvertent omission of transportation activities however, they are lower-bounded by their (positive) demands.

This instance of infeasibility is much simpler than would normally occur in practice. Still, the query commands and the TRACE command are often useful. There are additional commands for infeasibility tracing, such as RATEOF and REDUCE, which we shall demonstrate in the next chapter. Enter INTERP INFEAS to get some advice, especially when tackling more difficult cases. A collection of infeasible instances of a blending problem are in the MOREDATA library (see Appendix B).

4.4 A case of unboundedness.

Now we analyze an unbounded instance of WOODNET. To begin, we read in the packed LP file plus the syntax file with an execution file that was prepared for this exercise. This is shown in figure 4-19.

```
ANALYZE ... EXEC WOODUNBD
  WOODUNBD.....an unbounded network problem.
   Try to diagnose the cause of unboundedness. Use the
   EXPLAIN command to put your explanation into English.
```

Figure 4-19. Loading an Unbounded Case

One place to start is to look at the reduced costs of activities that are *non-optimal.* By this we mean the reduced cost for a nonbasic variable with Stat = Lower bound is negative, or for a nonbasic variable with Stat = Upper bound is positive (for minimization). The reason is that the solver found such an activity whose level can be increased without limit; and, the reduced cost is the net rate of change in the objective. This is shown in figure 4-20, using the ADDRIM command for the columns with activity levels at their lower bounds. The reduced cost range specification, D=-*/-.01, is the interval $(-\infty, -.01]$. The result shows that 20 columns have this property.

```
ANALYZE ... ADDR C * X EQ L D=-*/-.01
 Submatrix has 20 COLS

          SUM           MIN           MAX          AVERAGE
       ----------------------------------------------------
   L      30.000         0           20.000         1.500
   U   ...ALL VALUES ARE INFINITE
   C    - 810.000 -   1000.000       18.000      -  40.500
   X      30.000         0           20.000         1.500
   D  -  1163.000   - 925.000      - 1.0         -  58.150
       ----------------------------------------------------
```

Figure 4-20. Seeing the Reduced Cost Range Among Non-optimal Activities

In the absence of other information, we may as well start with an activity having the most negative reduced cost, –925, especially since it is much below the average. This is done in figure 4-21, and one column (namely, DMODE1) satisfies this. We next picture it with all of its rows.

```
ANALYZE ... PI *,* D=-925
                              Picture of submatrix
          D
          M
          O
          D
          E
          1
COST      - = MIN
DMODE     - =  0
```

Figure 4-21. Reducing the Search to the Activities with the Most Negative Price

We now want to trace through the matrix, adding other columns that balance the demand equation (DMODE), then their rows, and so on. The TRACE command does this, as shown in figure 4-22. This reveals a path from supply of mahogany in Seattle, transported for demand in Denver. The delivered cost is $75 ($55 supply cost + $20 transportation cost). Unlike the optimal instance, this version has a revenue for the demand activity of $1000. This gives a net cost of –$925 (equivalently, a revenue of $925), which is the reduced cost. If the path is not capacitated, this causes unboundedness because we can choose the levels of the 3 activities to be a common value that can increase without bound while satisfying the equations. (The constraint satisfaction is due to the equations being balanced by these 3 activities.) As this common value increases, the COST value tends to $-\infty$ at a rate of –925.

```
ANALYZE ... TRACE
 TRACE complete...submatrix is set
ANALYZE ... PIC
                                   Picture of submatrix
         D S T
         M M M
         O O O
         D S S
         E E E
         1 1 D
             E
COST     - + + = MIN
DMODE    -   + =  0
SMOSE      + - =  0
ANALYZE ... L
      MIN     COST    =-  1000 DMODE1 +  55 SMOSE1 +  20 TMOSEDE
      0     = DMODE   =- DMODE1 + TMOSEDE
      0     = SMOSE   =  SMOSE1 - TMOSEDE
```

Figure 4-22. Using the TRACE Command to Find a Complete Submatrix

It thus remains to verify that the path is uncapacitated. This is determined with the display of the columns, as shown in figure 4-23. Here we specified the parameter U (following //) to suppress the other fields of the display. You can use the EXPLAIN command, as before, to put the final analysis into English. Also, you might check the other demand activities having revenues (negative COST values).

```
ANALYZE ... D C //U
 3 Columns are in submatrix.
 COL      UP_BOUND
 -------------------
 DMODE1       *
 SMOSE1       *
 TMOSEDE      *
```

Figure 4-23. Displaying the Upper Bounds (only) of the Submatrix Columns to Verify the Path is Uncapacitated

This case of unboundedness was simplified by having a path structure, common in network models, and only one out-of-kilter non-optimal reduced cost. Our purpose is only to demonstrate the ANALYZE commands, keeping the analysis fairly simple. Additional commands, like RATEOF, are demonstrated in the next chapter, which offer additional computer-assistance for analysis.

Chapter 5. Further Examples

A model is to an analyst as a magnifying glass is to Sherlock Holmes —— it illuminates the clues.

In this chapter we show how to use some of the procedural commands of ANALYZE besides simple query and tracing. We begin with a continuation of analyses of the instances of the WOODNET model. To follow along enter EXEC WOODNET to read the optimal instance of the LP and syntax files into memory (c.f., figure 4-1). First, we demonstrate how to use the BLOCK command, including the analysis problem of tracing a flow of materials (both quantities and prices). Second, we show how to obtain and apply relevant *rates of substitution* with the RATEOF command. Third, we show how to get information from the basis, including redundancy detection. Fourth, we illustrate various ways to use the REDUCE command. Finally, we explain the AGGREGAT command and its application to aid infeasibility diagnosis.

5.1 Blocking.

The BLOCK command is a useful way to obtain macro views of the model instance. A *row block* is a set of rows, and a *column block* is a set of columns. By forming blocks in a variety of ways, we can picture how portions of the model relate to each other.

We saw that the SCHEMA command presents one way to get a syntax-based macro view of the model. We can also block by syntax, and we may do so on any submatrix, rather than the entire matrix. Figure 5-1 illustrates how to block by syntax with the result given for the WOODNET model. We begin by putting all rows and columns in the submatrix with the SUBMAT * command. Then, we specify SUB D R COST to delete the COST row, so the submatrix has the Supply and Demand rows (and all columns).

Next, we specify BLOCK ROW SYNTAX (abbreviated BL R S). This creates row blocks distinguished by the first letter of the name of each row. In the case of WOODNET, 2 row blocks are found. These are displayed with the BLOCK ROW DISPLAY (abbreviated BL R D). The display gives the block names, created by the blocking procedure, and the number of rows in each block. The mask, shown in the second column of the block display, is the intersection of the row names in the block. For example, all rows in block SYN.S match the name mask S.

```
ANALYZE ...SUB *
ANALYZE ...SUB D R COST
ANALYZE ...BL R S
           2 block(s) added
ANALYZE ...BL R D
2        ROW Blocks
Block   Mask    Number in Block
-------------------------------
SYN.D   D         20
SYN.S   S         15
```

Figure 5-1. Blocking Rows by Syntax

Advice from a student

> The asterisk (*) seems to be an overused symbol, relying on semantics for its meaning. We use it as a wild character in name masks, to represent infinity for numerical values, and now to indicate mixed coefficient signs in a block. Such is life in an ASCII world.

We then do likewise to block columns by syntax, resulting in the 3 column blocks displayed in figure 5-2. The BLOCK PICTURE (abbreviated BL P) shows the WOODNET matrix structure. The minus entry (-) in a *cell* (i.e., intersection of row and column blocks) indicates that all nonzero coefficients in that submatrix are negative. Similarly, the plus entry (+) in a cell indicates that all nonzero coefficients in that submatrix are positive. When the submatrix determined by a row and block contains both negative and positive coefficients, an asterisk (*) appears. A blank entry indicates all coefficient in that submatrix are zero.

```
ANALYZE ...BL C S
          3 block(s) added
ANALYZE ...BL C D
3         COL Blocks
Block    Mask    Number in Block
--------------------------------
SYN.T     T         55
SYN.S     S         18
SYN.D     D         20
ANALYZE ...BL P
               BLOCK PICTURE
        S S S
        Y Y Y
        N N N
        . . .
        D S T
SYN.D   -   +
SYN.S     + -
```

Figure 5-2. Blocking Columns by Syntax and Picturing Them with the Row Blocks

We may want to prepare this block picture for someone unfamiliar with this jargon, or we may want to spool the result for some documentation. In such cases the row and column block names are rather cryptic, so we rename them. Knowing the WOODNET syntax, we rename the rows and columns, as shown in figure 5-3. The first specification is `BLOCK ROW RENAME SYN.S SUPPLY` (command and keywords are abbreviated `BL R REN`). Then, as shown, the block picture appears more intelligible. In general, block names are limited to 8 characters; a longer name is truncated to the first 8 characters.

```
ANALYZE ...BL R REN SYN.S SUPPLY
ANALYZE ...BL R REN SYN.D DEMAND
ANALYZE ...BL C REN SYN.S SUPPLY
ANALYZE ...BL C REN SYN.D DEMAND
ANALYZE ...BL C REN SYN.T TRANSFER
ANALYZE ...BL P
                BLOCK PICTURE
            D S T
            E U R
            M P A
            A P N
            N L S
            D Y F
                E
                R
DEMAND      -   +
SUPPLY        + -
```

Figure 5-3. Renaming Blocks

Now we illustrate two things: how to find components of a submatrix, and how to query blocks by copying them to the submatrix.

Recall from figure 3-16 and the associated discussion that we can use the SYNTAX and TALLY commands to learn of the materials in the WOODNET model, coded into the names of the rows and columns. Figure 5-4 re-creates the setting by picturing the submatrix composed of all rows pertaining to mahogany – that is, that match the name mask `*MO` – and all columns whose associated activity level is at least 1. The resulting submatrix has the appropriate Supply and Demand balance rows and the relevant Demand, Supply and Transportation activities.

After picturing the submatrix, the current blocks are erased, where we specify:

```
BL  R  E   *
|   |  |   |
|   |  |   └──► all blocks
|   |  └──► ERASE
|   └──► ROW
└──► BLOCK
```

This need to erase the blocks reflects the ANALYZE requirement:

Each row and column can be in at most one block.

Then, we enter the BLOCK command (without specifications) to obtain the components of the submatrix.

```
ANALYZE ... PI *MO,* X=1/*
                        Picture of submatrix
         D D D D D S S T T T T T
         M M M M M M M M M M M M
         0 0 0 0 0 0 0 0 0 0 0 0
         C D L S S S S S S S S S
         H E A E F E F E E E F F
         1 1 1 1 1 1 1 C D S L S
                       H E F A E
DMOCH    -             +         =  0
DMODE      -             +       =  0
DMOLA        -               +   =  0
DMOSE          -               + =  0
DMOSF            -         +     =  0
SMOSE              +   - - -     =  0
SMOSF                +       - - =  0
ANALYZE ... BL R E *
ANALYZE ... BL C E *
ANALYZE ... BLOCK
          2 components found
```

Figure 5-4. Picturing a Flow of Material and Blocking the Resulting Submatrix into its (2) Components

In this case, 2 components are found. Each component must represent flows of mahogany from supplies to demands. Before proceeding, notice that both row blocks and column blocks were erased. If we erased only the row blocks, we can get more components, contrary to what we want. This is because columns that are already in a block are ignored when components are sought. This feature becomes important when we seek a block-link structure, which we shall do in chapter 7. For now, just know that it is necessary to erase all blocks to get what we want in this exercise.

We now look at these disjoint flows separately and explain them. To separate the flows, we copy each component to the submatrix where we can then query it. This is shown in figure 5-5, where we need to know that the BLOCK command results in component blocks named COMP.001, COMP.002, ... for the row and column component strips. After the block copies are done, we picture the submatrix of this component.

```
ANALYZE ... SUBMAT CL
 Submatrix has  0 rows,  0 columns, and 0 nonzeroes
 SUBMAT (enter ? for menu)...
      (this line is blank...a return was entered from the keyboard)
ANALYZE ... BLOCK ROW COPY FROM COMP.001
ANALYZE ... BLOCK COL COPY FROM COMP.001
ANALYZE ... PI
                          Picture of submatrix
        D D D S T T T
        M M M M M M M
        O O O O O O O
        C D S S S S S
        H E F E E E E
        1 1 1 1 C D S
                H E F
DMOCH   -       +     =  0
DMODE     -       +   =  0
DMOSF       -       + =  0
SMOSE         + - - - =  0
```

Figure 5-5. Copying and Picturing the First Component

Let us pause to observe this relationship between blocks and the submatrix. Blocks are designed to capture various macro views and may be created by procedures, such as finding components, which may be part of an analysis or used as a check for model management. The submatrix, on the other hand, is designed to provide flexible query capability. Communication between the two gives an integrated system that does not require separate query commands for blocks. This is the function of the BLOCK COPY either FROM a block to the submatrix or TO a block from the submatrix. Now let us resume our analysis.

After seeing the component, we may proceed in several ways. In figure 5-6, the SUBMAT command was first used (directly) to add the COST row. Then, the LIST command, using the EQUATION format, gives an algebraic view of this submatrix.

```
ANALYZE ... SUB A R COST
ANALYZE ... L
      MIN    COST    =   55 SMOSE1 +  18 TMOSECH +  20 TMOSEDE
                      +  10 TMOSESF
      0    = DMOCH   =- DMOCH1 + TMOSECH
      0    = DMODE   =- DMODE1 + TMOSEDE
      0    = DMOSF   =- DMOSF1 + TMOSESF
      0    = SMOSE   =  SMOSE1 - TMOSECH - TMOSEDE - TMOSESF
```

Figure 5-6. Listing the Submatrix Equations

At this point it seems timely to understand what the rows and columns mean. We do this, as shown in figure 5-7, with the EXPLAIN command.

```
ANALYZE ... EXP R
    Row COST is the objective (MINIMIZE).
    Row DMOCH balances demand of mahogany in Chicago.
    Row DMODE balances demand of mahogany in Denver.
    Row DMOSF balances demand of mahogany in San Francisco.
    Row SMOSE balances supply of mahogany in Seattle.
ANALYZE ... EXP C
    Column DMOCH1 demands mahogany at Chicago.
    Column DMODE1 demands mahogany at Denver.
    Column DMOSF1 demands mahogany at San Francisco.
    Column SMOSE1 supplies mahogany at Seattle.
    Column TMOSECH transports mahogany from Seattle to Chicago.
    Column TMOSEDE transports mahogany from Seattle to Denver.
    Column TMOSESF transports mahogany from Seattle to San Francisco.
```

Figure 5-7. Explaining the Rows and Columns in the First Component

From the information in figures 5-6 and 5-7 we can put most of the story together, at least for the Seattle supplier associated with the first component. It is useful, however, to complete this part of the analysis by seeing the levels of flow and capacities that may be binding. Figure 5-8 uses the DISPLAY command to show these.

```
ANALYZE ... D R
5 Rows are in submatrix.
ROW      STAT  LEVEL (Y)     LO_BOUND      UP_BOUND      PRICE (P)
-------------------------------------------------------------------
COST      B      25135.000      - *            *            - 1.0
DMOCH     L          0            0            0            73.000
DMODE     L          0            0            0            75.000
DMOSF     L          0            0            0            65.000
SMOSE     L          0            0            0            55.000
ANALYZE ... D C
7 Columns are in submatrix.
COL      STAT  LEVEL (X)     LO_BOUND      UP_BOUND      PRICE (D)
-------------------------------------------------------------------
DMOCH1    L         25.000       25.000        25.000       73.000
DMODE1    L         20.000       20.000        20.000       75.000
DMOSF1    L         30.000       30.000        30.000       65.000
SMOSE1    B         75.000       0              *            0
TMOSECH   B         25.000       0              *            0
TMOSEDE   B         20.000       0              *            0
TMOSESF   B         30.000       0              *            0
```

Figure 5-8. Displaying Rim Elements of the Rows and Columns in the First Component

With the above information we can explain the flow of mahogany from the Seattle supplier as follows.

```
Seattle produces 75 units @$55, which is not at capacity.  These
are shipped to 3 consumers located in Chicago, Denver and San
Francisco.  The Chicago shipment is 25 units with a
transportation cost of $18, so its consumer price is $73.  The
Denver shipment is 20 units with a transportation cost of $20,
so its consumer price is $75.  The San Francisco shipment is 30
units with a transportation cost of $10, so its consumer price
is $65.  None of the shipments are limited by capacity
constraints, which is why each consumer's price is simply the
delivered cost.
```

This text explanation is what the automatic interpretation did (c.f., figure 4-15). Another presentation is graphical, as shown in figure 5-9. The information we have gathered from the BLOCK and query commands were used to construct this presentation.

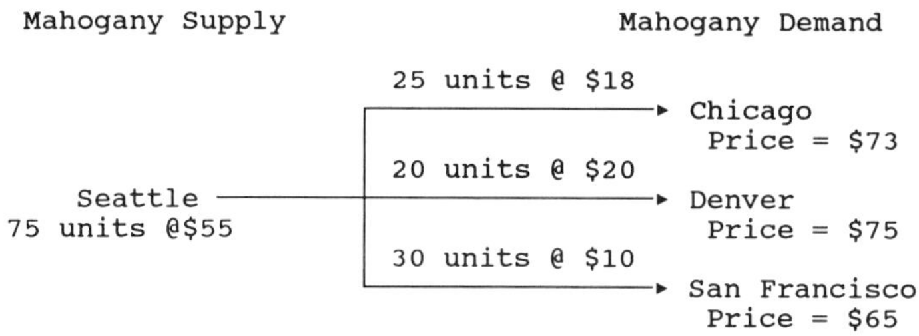

Figure 5-9. A Graphical Depiction of the First Component

Now let us do the analysis for the second component. Proceeding as before, we clear the submatrix and copy the second blocks of rows and columns to the submatrix, where we then picture the flow. This is shown in figure 5-10.

```
ANALYZE ... SUBMAT CL
 Submatrix has  0 rows,  0 columns, and  0 nonzeroes
 SUBMAT (enter ? for menu)...
     (this line is blank...a return was entered from the keyboard)
ANALYZE ... BL R C F COMP.002
ANALYZE ... BL C C F COMP.002
ANALYZE ... PI
                                   Picture of submatrix
         D D S T T
         M M M M M
         O O O O O
         L S S S S
         A C C F F
         1 1 1 L S
               A E
DMOLA    -     +   =  0
DMOSE      -     + =  0
SMOSF        + - - =  0
```

Figure 5-10. Copying and Picturing the Second Component

We want to see the costs, so we add the COST row to the submatrix with the direct use of the SUBMAT command before listing (in EQUATION format), as shown in figure 5-11.

```
ANALYZE ... SUB A R COST
ANALYZE ... L
      MIN     COST     =   45 SMOSF1 +  5 TMOSFLA +  8 TMOSFSE
      0     = DMOLA    =- DMOLA1 + TMOSFLA
      0     = DMOSE    =- DMOSE1 + TMOSFSE
      0     = SMOSF    =  SMOSF1 - TMOSFLA - TMOSFSE
```

Figure 5-11. Adding the COST Row Before Listing the Equations

It seems we have a similar story, except this supplier sends to only 2 consumers. The EXPLAIN command can tell us who "this supplier" and the "2 consumers" are. This is shown in figure 5-12.

```
ANALYZE ... EXP R
    Row COST is the objective (MINIMIZE).
    Row DMOLA balances demand of mahogany in Los Angeles.
    Row DMOSE balances demand of mahogany in Seattle.
    Row SMOSF balances supply of mahogany in San Francisco.
ANALYZE ... EXP C
    Column DMOLA1 demands mahogany at Los Angeles.
    Column DMOSE1 demands mahogany at Seattle.
    Column SMOSF1 supplies mahogany at San Francisco.
    Column TMOSFLA transports mahogany from San Francisco to Los Angeles.
    Column TMOSFSE transports mahogany from San Francisco to Seattle.
```

Figure 5-12. Explaining the Rows and Columns in the Second Component

Finally, to get a similar story we display the solution and rim values to get quantities of flow and capacity information, as shown in figure 5-13. There is, however, an important difference from the previous case: the supply activity (SMOSF1) is at its capacity, causing the consumer prices not to be equal to their delivered costs!

```
ANALYZE ... SUB D R COST
ANALYZE ... D R
 3 Rows are in submatrix.
 ROW      STAT  LEVEL (Y)     LO_BOUND      UP_BOUND       PRICE (P)
 ------------------------------------------------------------------
 DMOLA     L          0             0             0           68.000
 DMOSE     L          0             0             0           71.000
 SMOSF     L          0             0             0           63.000
ANALYZE ... D C
 5 Columns are in submatrix.
 COL      STAT  LEVEL (X)     LO_BOUND      UP_BOUND       PRICE (D)
 ------------------------------------------------------------------
 DMOLA1    L         20.000        20.000        20.000       68.000
 DMOSE1    L         10.000        10.000        10.000       71.000
 SMOSF1    U         30.000         0            30.000    -  18.000
 TMOSFLA   B         20.000         0             *            0
 TMOSFSE   B         10.000         0             *            0
```

Figure 5-13. Displaying Rim Elements of the Rows and Columns in the Second Component

The San Francisco supplier, of the second component, has a reduced cost on the associated activity – that is, `Stat=U` for column `SMOSF1` to indicate its level is equal to its upper bound (30 units), and we see its dual `Price` = –18. This adds to the supply cost, which requires further analysis (c.f., figure 4-16).

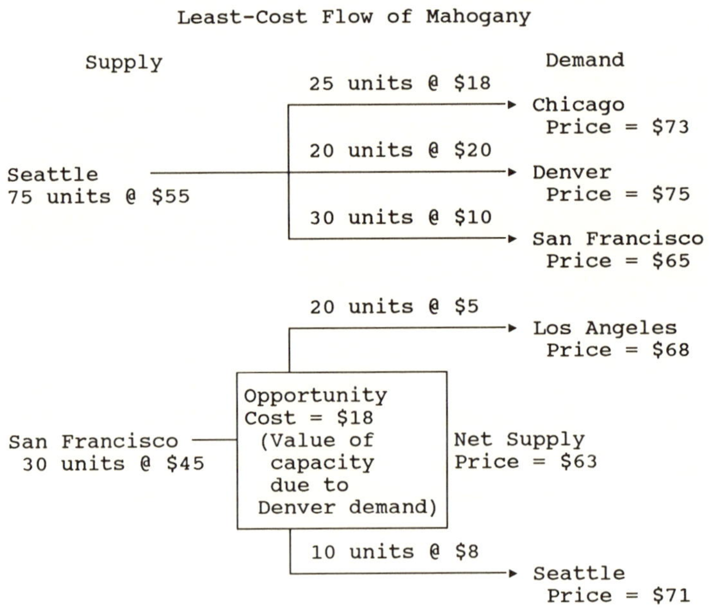

Figure 5-14. A Final Depiction of the Analysis

The BLOCK command also finds embedded structures, such as a network. Figure 5-15 shows its application to WOODNET. After putting all rows and columns in the submatrix, all blocks are erased. Then, the BLOCK NETFORM (abbreviated `BL N`) is specified. This results in the creation of 3 column blocks, which are displayed. The block named NET.PROD contains *production activities* – that is, those with exactly one nonzero coefficient in the matrix (not counting the objective row), which is positive. The block named NET.CONS contains *consumption activities* – that is, those with exactly one nonzero coefficient in the matrix, which is negative. The block named NET.TRAN contains *transfer activities* – that is, those with exactly one negative coefficient and exactly one positive coefficient in the matrix.

```
ANALYZE ...SUB *
ANALYZE ...BL R E *
ANALYZE ...BL C E *
ANALYZE ...BL N
 NET.TRAN has         55 activities
 NET.CONS has         20 activities
 NET.PROD has         18 activities
ANALYZE ...BL C DISP
 3         COL Blocks
 Block    Mask    Number in Block
 -------------------------------
 NET.TRAN T        55
 NET.CONS D****1   20
 NET.PROD S        18
```

Figure 5-15. Blocking a Maximal Embedded Netform

After blocking the columns into Production, Consumption and Transfer activities, the rows are blocked by syntax. The resulting block picture, shown in figure 5-16, includes the COST row as the only member of row block SYN.C.

```
ANALYZE ...BL R S
          3 block(s) added
ANALYZE ...BL P
             BLOCK PICTURE
        N N N
        E E E
        T T T
        . . .
        P C T
        R O R
        O N A
        D S N
SYN.C   +   +
SYN.D     - +
SYN.S   +   -
```

Figure 5-16. Picturing the Netform Over the Row Syntax

In this model all columns entered one of the netform blocks because WOODNET is a network model. In general, the BLOCK NETFORM specification finds a *maximal* embedded netform. This means no column could be added to the blocks that cannot be classified as one that consists of a single nonzero (Production or Consumption) or an oppositely signed pair of nonzeroes (Transfer) over the body of the submatrix (i.e., ignoring free rows, like the objective). If the model has only inequalities, there is no ambiguity for the signing. In that case, we have the equivalent of the canonical form `Ax ≥ b` (for minimization), so there is no confusion about supplies versus demands. When the model contains equations, however, one must take care in using this (and other) procedures that depend upon the signs of the coefficients.

Because the blocking applies only to the submatrix, one may look for special structures, combining different blocking functions. For example, we can combine the netform blocking with component blocking to find a *block-link* embedded structure. (In general, one may think of the blocks as regional production processes, and the links as activities to move materials around.) Figure 5-17 illustrates this with a small model, called BLOCKNET, constructed just for testing and illustration.

```
ANALYZE ...  READ M BLOCKNET
NAME   BLOCKNET
         6 rows...                1 free
         7 columns...               14 nonzeroes
 RHS section complete...NAME = RHS
 BOUNDS complete...NAME = BNDSET
 Matrix read in
 Initialized solution to all logical basis.
ANALYZE ...  PIC *,*
                                Picture of submatrix
         C C C C C N N
         1 1 2 2 2 1 2
         1 2 1 2 3 2 1
R11      + +       -   = 0
R12      + +         - = 0
R21          +   + +   = 0
R22          + +       = 0
R23            + +   + = 0
```

Figure 5-17. Reading a Matrix With a Block-Link Structure

We first read the matrix file into memory. Then, we picture the entire matrix to show its structure and put all rows and columns in the submatrix. (Whenever a new LP is read in, such as by reading a matrix file, it replaces the old one. The submatrix and blocks are cleared, but session controls, such as screen settings, remain unchanged.)

We begin with blocking the network columns, as shown in figure 5-18. In this case there are no production or consumption activities, and the transfer block is displayed. As shown in the subsequent picture, the columns in this block are N12 and N21. They represent transfer activities, and there are no production or consumption activities (so blocks NET.PROD and NET.CONS were not created). This completes the first of two steps in finding a block-link structure.

```
ANALYZE ...  BL N
 NET.TRAN  has      2 activities
 NO CONSUMPTION BLOCK
 NO PRODUCTION BLOCK
ANALYZE ...  BL C D
 1         COL Block
 Block    Mask    Number in Block
 ------------------------------
 NET.TRAN N        2
```

Figure 5-18. Finding the Links in a Block-Link Structure

Now we block for components, as shown in figure 5-19, which demonstrates a consequence of the 1 block limit. Because the network columns are already in a block (namely, NET.TRAN), they are ignored. Then, the remaining model has 2 components, as shown.

```
ANALYZE ...BL
           2 components found
ANALYZE ...BL P
              BLOCK PICTURE
        N C C
        E O O
        T M M
        . P P
        T . .
        R 0 0
        A 0 0
        N 1 2
COMP.001 - +
COMP.002 +   +
```

Figure 5-19. Finding the Blocks in a Block-Link Structure

A final step, shown in figure 5-20, is to verify that the block-link structure we found agrees with what we can see by inspection in figure 5-17. That is, block 1 has rows R11 and R12 and columns C11 and C12; block 2 has rows R21, R22 and R23 and columns C21, C22 and C23.

```
ANALYZE ...  BL R D
 2         ROW Blocks
 Block    Mask    Number in Block
 --------------------------------
 COMP.001 R1          2
 COMP.002 R2          3
ANALYZE ...  BL C D
 3         COL Blocks
 Block    Mask    Number in Block
 --------------------------------
 COMP.002 C2          3
 COMP.001 C1          2
 NET.TRAN N           2
```

Figure 5-20. Verifying the Block-Link Structure

The BLOCK command has a variety of uses, and we have demonstrated some of them. You can also create a block of arbitrary rows or columns by first creating the submatrix. Then, specify BLOCK ROW COPY TO *blockname* (or simply BL R C T *blockname*) to create the row block. In general, the block copy (ROW or

COLUMN) can be TO a block from the submatrix; or, it can be FROM a block to the submatrix. The former allows blocks to be created for reasons other than the automatic blocking. The latter enables query of the block, since it enters the submatrix.

5.2 Rates of substitution.

In this section we illustrate the use of the RATEOF command. First, let us define some terms and concepts. The resident *Basic* solution represents a re-write of the original equations, $y = Ax$, to an equivalent system that puts all basic variables on the left side of the equation and all nonbasic variables on the right. As a first step, partition A as follows.

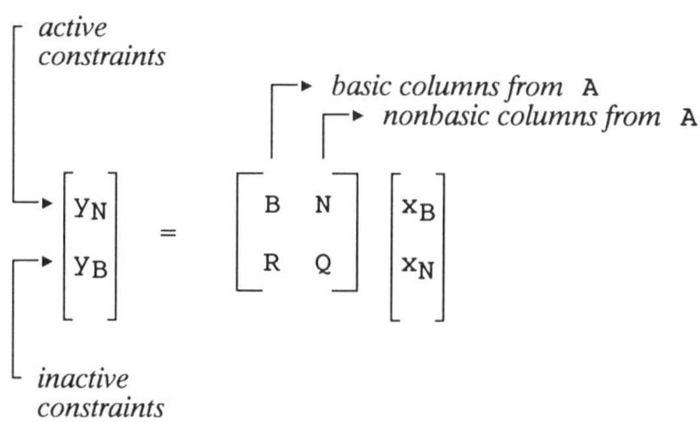

Then, after algebraic manipulation, the equivalent system is given by:

$$\begin{bmatrix} x_B \\ y_B \end{bmatrix} = \begin{bmatrix} B^{-1} & -B^{-1}N \\ RB^{-1} & Q-RB^{-1}N \end{bmatrix} \begin{bmatrix} y_N \\ x_N \end{bmatrix}$$

where $[x_B, y_B]^T$ is the vector of basic activity levels, and $[y_N, x_N]^T$ is the vector of nonbasic activity levels. At this point the nonbasic levels are at one of their bounds (L or U), and this uniquely determines the basic levels.

The basis inverse transforms the linear system of equations, so we can see how the basic variables must change if we impose a change in any of the nonbasic levels, such as by changing its bound value. The transformed matrix coefficients thus describe *rates of substitution*. This is what the RATEOF command computes and displays: the rate of each basic variable in the submatrix with respect to the nonbasic variable specified. The rate at which the i-th basic level changes with respect to a (forced) change in the j-th nonbasic level is the (i, j) entry in the (partitioned) matrix that defines the basic equations.

To illustrate, we begin with a small example so that we can see the arithmetic. As shown in figure 5-21, we first read the packed file RATETST.PCK into memory (this was created when you executed ATEST), then we list its equations.

```
ANALYZE ... REA P RATETST
 Packed LP read from RATETST.PCK

ANALYZE ... L *,*
      MIN     OBJ       =- X1
   15      >= R1        =  X2 +   2 X1
   15      <= R2        =  X1 +   2 X2
```

Figure 5-21. Reading a Small Problem To See Rates of Substitution

The equations may be put into matrix form:

$$\begin{bmatrix} OBJ \\ R1 \\ R2 \end{bmatrix} = \begin{bmatrix} -1 & 0 \\ 2 & 1 \\ 1 & 2 \end{bmatrix} \begin{bmatrix} X1 \\ X2 \end{bmatrix}$$

Inverting the lower 2×2 matrix (B), this is re-written as:

$$\begin{matrix} x_B \to \\ x_B \to \\ y_B \to \end{matrix} \begin{bmatrix} X1 \\ X2 \\ OBJ \end{bmatrix} = \begin{bmatrix} 2/3 & -1/3 \\ -1/3 & 2/3 \\ -2/3 & 1/3 \end{bmatrix} \begin{matrix} \\ \\ \leftarrow \text{Row Prices} \end{matrix} \begin{bmatrix} R1 \\ R2 \end{bmatrix} \leftarrow y_N$$

Now let us use the RATEOF command to see these values. First, in figure 5-22, we see the rates of substitution for the basic columns, `X1` and `X2`. Note the entries agree (approximately) with the above. The *Least Term* and *Greatest Term* mean how much each nonbasic variable (`R1` and `R2`) contributes to the overall range of the basic variable when considering only the one basic equation with X1. For example, consider the meaning of the basic equation for X1. We have:

$$X1 = 5 + 0.6666667\ \delta R1 - 0.3333333\ \delta R2$$

where δR1 denotes a change in the level of R1, and δR2 denotes a change in the level of R2. From figure 5-22 we know

$$-\infty \le R1 \le 15 \text{ and } 15 \le R2 \le \infty.$$

The least value of X1, therefore, is when the level of R1 is –∞ and the level of R2 is +∞. The greatest value of X1 is when the level of R1 is 15 and the level of R2 is 15. The first term after the current level (5) in the above equation, therefore, contributes –∞ to the least range value and 10 to its greatest range value. (The 10 is the product of the rate and the bound value of R1: 15×0.6666667.) Similarly, the contributions of R2 to the range of X1 are –∞ and –5, as shown. The X1 range is the sum: (–∞, 5].

```
ANALYZE ... RAT C *
  Rates of substitution for COL X1
  ...Status=B Level=5.000  Price=0
 Nonbasic    Rate of       Least        Greatest
 Variable  Substitution    Term         Term
 ================================================
 R1          0.6666667       - *          10.000
 R2        - 0.3333333       - *        -  5.000
 ------------------------------------------------
       X1       Range:       - *           5.000

  Rates of substitution for COL X2
  ...Status=B Level=5.000  Price=0
 Nonbasic    Rate of       Least        Greatest
 Variable  Substitution    Term         Term
 ================================================
 R1        - 0.3333333    -   5.000        *
 R2          0.6666667       10.000        *
 ------------------------------------------------
       X2       Range:        5.000        *
```

Figure 5-22. Rates of Substitution for the (Basic) Columns

You might now specify RAT //E to see the rates in equation format. It will appear like an equation listing, except that the change symbol (δ) is omitted. There are occasions, however, when we want to know which nonbasic variables have nonzero rates of substitution with some basic variables. Figure 5-23 illustrates this for the row variables, R1 and R2. Before we discuss them, consider the rates for the objective row (OBJ), which is always basic because it is a free row (i.e., it has no bound). The rates agree with the above matrix equation, and the least value in its range must be the minimum value if the basic solution is optimal.

Now examine the OBJ row for each of the nonbasic row variables. The rates of substitution must equal the price of the row. As we shall see below, the reduced cost of a column must also equal its rate of substitution for the objective row.

The *Least Change* and *Greatest Change* are the ranges imposed by the bounds of the basic variables on the level of the nonbasic variable. For example, consider R1. Keeping the level of R2 unchanged, the basic equations describe how the basic levels change with a change in the level of R1:

$$0 \le X1 = 5 + 0.6666667\ \delta R1$$
$$0 \le X2 = 5 - 0.3333333\ \delta R1$$

The first inequality imposes a lower bound of –7.5 on the change (δR1), and the second inequality imposes an upper bound of 15, as shown. The range of the level of R1 is its current value plus the range of its change. Since the current level of R1 is 15, its range is [15–7.5, 15+15], which is [7.5, 30], as shown.

Now consider the optimal instance of WOODNET, read into memory (with its syntax file) by EXEC WOODNET (c.f., figure 4-1). You may also just enter READ P WOODNET since we shall not use the syntax here (but recall the explanations given in the previous chapter).

```
ANALYZE ... RAT R *
  Rates of substitution for ROW OBJ
  ...Status=B Level=-5.000  Price=-1.0
 Nonbasic    Rate of       Least        Greatest
 Variable  Substitution    Term         Term
 =================================================
 R1        - 0.6666667    -  10.000        *
 R2          0.3333333        5.000        *
 -------------------------------------------------
       OBJ       Range:   -   5.000        *

  Rates of substitution for ROW R1
  ...Status=U Level=15.000  Price=-0.6666700
 Basic       Rate of       Least        Greatest
 Variable  Substitution    Change       Change
 =================================================
 OBJ       - 0.6666667      - *            *
 X1          0.6666667    -   7.500        *
 X2        - 0.3333333      - *         15.000
 -------------------------------------------------
       R1        Range:       7.500     30.000
       Blockage:            X1          X2

  Rates of substitution for ROW R2
  ...Status=L Level=15.000  Price=0.3333300
 Basic       Rate of       Least        Greatest
 Variable  Substitution    Change       Change
 =================================================
 OBJ         0.3333333      - *            *
 X1        - 0.3333333      - *         15.000
 X2          0.6666667    -   7.500        *
 -------------------------------------------------
       R2        Range:       7.500     30.000
       Blockage:            X2          X1
```

Figure 5-23. Rates of Substitution for the Rows

Figure 5-24 begins with the rates of substitution for the nonbasic demand column we analyzed in the previous chapter (namely, DMOCH1).

```
ANALYZE ... RATEOF COL DMOCH
  Rates of substitution for COL DMOCH1
  ...Status=L Level=25.000  Price=73.000
 Basic        Rate of        Least          Greatest
 Variable  Substitution     Change          Change
 ==================================================
 COST              73.000       - *             *
 TMOSECH            1.0       -  25.000         *
 SMOSE1             1.0       -  75.000         *
 --------------------------------------------------
       DMOCH1   Range:            0             *
       Blockage:             TMOSECH      (no block)
```

Figure 5-24. Rates of Substitution for the Nonbasic Demand Activity DMOCH1

The specification of the column name omitted the last character (1) because this is the only column whose name begins with DMOCH – that is, trailing blanks are the same as mask characters. Of course, if there was also a DMOCH2, we would get the rates for both of them unless we specified the full name, DMOCH1.

The display begins with the activity's status (L), its level (25), and its Price (73). Then, the basic variables are tabulated. The transformed matrix entry of the COST row is precisely the reduced cost (73). This means that if we force a marginal change in the level of DMOCH1, the COST will change at the rate of $73 per unit of change in the demand. Because the objective row is defined to be free, it cannot block the level of the nonbasic activity - that is, because the bounds on COST are infinite, the nonbasic level is not *blocked* by any limit due to maintaining feasibility.

The supply activity SMOSE1 and the transportation activity TMOSECH have rates of substitution = 1. This means that for every unit we change the level of DMODE1, these 2 activities each change by exactly one unit as well, and the direction of change is the same. Knowing the model's structure, we see that if demand is increased, the supply and transportation activities linked to this demand must increase the same amount for the levels to continue to satisfy the equations.

If we were to decrease the demand from its current level of 20, the supply and transportation activity levels must also decrease. The transportation level falls to 0 when the demand level reaches 0, at which point the demand cannot be driven lower without forcing the level of the transportation activity to become negative. This therefore imposes a least value of 0, as shown in the DMOCH1 Range. Because none of the basic variables have upper bounds, the demand level may be increased while satisfying the equations at the displayed rates of substitution - that is, by increasing the supply and transportation activity levels at the same rate as the demand level increases.

For price analysis it is useful to add the objective row to the display, so we can see *how* the level of COST increases if we increase this demand. This is done with the `//C` (or `//OBJECTIV`) parameter specification, as shown in figure 5-25. (It is not necessary to re-specify `COLUMN DMOCH1` since that is the submatrix setting.) From this display we can account for the net increase in COST of $73. By convention, we set the objective coefficient = −1 for minimization, which is also displayed since the objective row is always basic.

```
ANALYZE ... RATE //C
  Rates of substitution for COL DMOCH1
  ...Status=L Level=25.000  Price=73.000
 Basic        COST          Rate of        Least          Greatest
 Variable  Coefficient  Substitution    Change         Change
============================================================
 COST              - 1.0          73.000       - *             *
 TMOSECH            18.000         1.0      -  25.000          *
 SMOSE1             55.000         1.0      -  75.000          *
 -----------------------------------------------------------
                    DMOCH1   Range:            0               *
                    Blockage:               TMOSECH     (no block)
```

Figure 5-25. Adding the Objective Coefficients of the Basic Variables to the Rate Display

For flow analysis it is useful to add the basic levels to the display. This is done with the `//X` (or `//LEVEL`) specification, as shown in figure 5-26. (We abbreviated the RATEOF command as RA.) The absence of a row or column conditional retains the submatrix, so the table is the same as above except for the special column that shows the level of each basic activity tabulated – that is, for which the rate of substitution is not zero.

```
ANALYZE ... RA //X
  Rates of substitution for COL DMOCH1
  ...Status=L Level=25.000  Price=73.000
 Basic          Basic        Rate of       Least          Greatest
 Variable       Level      Substitution    Change         Change
==============================================================
 COST         25135.000        73.000         - *              *
 TMOSECH         25.000         1.0        -  25.000           *
 SMOSE1          75.000         1.0        -  75.000           *
--------------------------------------------------------------
                    DMOCH1    Range:            0              *
                    Blockage:              TMOSECH      (no block)
```

Figure 5-26. Adding the Levels of the Basic Variables to the Rate Display

Now recall the second price analysis in the previous chapter, where we had a degeneracy to explain why the price of demand in Los Angeles was greater than its reduced cost. The analysis brought us to the demand activity DMOLA1. In figure 5-27 we proceed as above. We display its rates of substitution with the basic levels. This is done by the `RAT C DMOLA1 //X` specification, where the column conditional causes the submatrix to change appropriately.

```
ANALYZE ... RAT C DMOLA //X
  Rates of substitution for COL DMOLA1
  ...Status=L Level=20.000  Price=68.000
 Basic          Basic        Rate of       Least          Greatest
 Variable       Level      Substitution    Change         Change
==============================================================
 COST         25135.000        68.000         - *              *
 TMOSEDE         20.000         1.0        -  20.000           *
 TMOSFLA         20.000         1.0        -  20.000           *
 SMOSE1          75.000         1.0        -  75.000           *
 TMOSFDE          0           - 1.0           - *              0
--------------------------------------------------------------
                    DMOLA1    Range:            0           20.000
                    Blockage:              TMOSEDE        TMOSFDE
```

Figure 5-27. Displaying the Rates of Substitution for the Nonbasic Activity DMOLA1 with Basic Levels

The least level of DMOLA1 is when the demand is driven to 0. Recall the delivery path is composed of activities SMOSF1 and TMOSFLA, but only the latter (transportation) activity is basic; the former (supply) activity is non-basic at its upper bound, so it definitionally cannot change its level for marginal sensitivity analysis. If the demand decreases, the level of TMOSFLA decreases at the unit rate of substitution shown. The negative rate of substitution for the basic activity TMOSFDE reflects the displacement of its delivery by the freed supply. This is how we explained the demand price of $68 in LA, and seeing the rates of substitution offers another view of the situation.

Here we also see the degeneracy imposing a limit of 0 on the greatest change of level of activity DMOLA1. This means this particular basic solution cannot provide the marginal price associated with *increasing* the demand for mahogany in LA; the $68 is the value of *decreasing* this demand level because the basic solution is compatible with the displacement events that occur when this demand is decreased. The *consumer's price*, therefore, is a bit more complicated than we thought. The $68 is only a *one-sided* rate (i.e., a left derivative), associated with the decreasing change in LA demand. It is a bound for increasing the demand – that is, the total MIN COST increases at a rate of *at least* $68 per unit increase in the demand for mahogany in LA. To get the exact rate, the basis must be changed to an alternative optimal solution that contains the delivery path to accommodate the perturbation.

It is this one-sided property that the rates of substitution reveal better than the simple query methods we used before. It also sharpens our explanation if the notion of *consumer's price* is important for the final analysis.

Now consider the unbounded instance of WOODNET, which you can bring in by entering EXEC WOODUNBD (c.f., figure 4-19) or by entering READ P WOODUNBD (since we shall not use the syntax file). Recall that in §4.4 we homed in on the nonbasic activity `DMODE1`, so let's look at its rates of substitution in figure 5-28.

```
ANALYZE ... RAT C DMODE //C
  Rates of substitution for COL DMODE1
  ...Status=L Level=20.000  Price=-925.000
 Basic       COST         Rate of        Least          Greatest
 Variable  Coefficient  Substitution    Change         Change
 ==============================================================
 COST            - 1.0      - 925.000      - *              *
 TMOSEDE         20.000        1.0       -  20.000          *
 SMOSE1          55.000        1.0       -  75.000          *
 --------------------------------------------------------------
                   DMODE1   Range:           0              *
                   Blockage:             TMOSEDE       (no block)
```

Figure 5-28. Showing the Rates of Substitution for the Activity That Indicated Unboundedness

As before, the response begins with the activity's Status (L), its Level (20), and its Price (–925). Then, the Basic Variables are tabulated. Since the //C parameter was specified, the second column shown is the COST coefficient of each basic variable. The transformed matrix entry is precisely the reduced cost (–925) because that is the rate of substitution of the level of the COST row with respect to the level of the activity DMODE1.

The supply activity SMOSE1 and the transportation activity TMOSEDE have rates of substitution = 1. As before, this means that for every unit we change the level of DMODE1 these two activities each change by exactly one unit as well, and the direction of change is the same. Knowing the model's structure, we see that if demand is increased, the levels continue to satisfy the equations if, and only if, the linked supply and transportation activity levels increase by the same amount. The costs shown (55 and 20, respectively) give the net rates of change in the total cost ($75 = $55 supply + $20 transportation), which offset the original COST value of DMODE1 = –1000 to yield the rate in the COST row of the table. Theoretically, this is always the reduced cost of the nonbasic activity, but there may be some round-off errors for models that are not pure networks.

If we were to decrease the demand from its current level of 20, the supply and transportation activity levels must also decrease. The transportation level falls to 0 when the demand level reaches 0, at which point the demand cannot be driven lower without forcing the transportation activity's level to become negative. This therefore imposes a least level of 0, as shown in the DMODE1 range at the bottom of the display. The relevant case is that there are no upper bounds on these basic activities, so this tells us that the nonbasic activity's level can become indefinitely great without violating the equations.

Try some of the other options. In particular, enter RATEOF //EQ to see equation format. Enter RATEOF //EQ OBJ to see equation format with objective rates added. Enter RATEOF //SYNTAX to give syntax-based translations, and enter RATEOF //SYN,COL to include only the columns (excluding rows in the output).

The variable chosen to induce a change is called the *impulse variable*, and those variables affected are called *response variables*. So far, we have specified nonbasic impulse variables in the RATEOF command. There are times, however, when we want to see all nonbasic variables that affect a particular basic variable. Algebraically, we are asking for the re-write of the equations, relative to the basis. For a particular basic variable, say x_β, the basic equations have the form:

$$x_\beta = X_\beta + \Sigma_j \; R(x_\beta, x_j)\,\delta x_j$$

- δx_j → *change in the j-th nonbasic level*
- $R(x_\beta, x_j)$ → *rate of substitution*
- X_β → *current level of* x_β

To illustrate, first READIN P WOODNET to bring the optimal instance of WOODNET into memory. Then, figure 5-29 illustrates this for activity TMOSEDE, where the nonbasic levels are displayed as well. Now the rate shown is the same as before; however, unlike the case where the impulse is a nonbasic variable affecting all basic variables, these responses are not simultaneous ones. They express the effect of each nonbasic variable on this basic variable. For example, we see that an increase in the level of row DMODE causes an increase by the same amount in TMOSEDE – that is, its rate of substitution is 1.

```
ANALYZE ... RAT C TMOSEDE //X
  Rates of substitution for COL TMOSEDE
  ...Status=B Level=20.000  Price=0
 Nonbasic      Nonbasic     Rate of       Least         Greatest
 Variable      Level      Substitution    Term          Term
===============================================================
 DMODE            0             1.0          0             0
 DMOLA            0             1.0          0             0
 DMOSE            0             1.0          0             0
 SMOSF            0             1.0          0             0
 DMODE1         20.000          1.0        20.000        20.000
 DMOLA1         20.000          1.0        20.000        20.000
 DMOSE1         10.000          1.0        10.000        10.000
 SMOSF1         30.000        - 1.0      - 30.000          0
 TMOSELA          0           - 1.0        - *             0
 TMOSFCH          0             1.0          0             *
 ---------------------------------------------------------------
                   TMOSEDE  Range:         - *             *
```

Figure 5-29. Specifying a Basic Variable as the Impulse to Obtain Nonbasic Response Variables that Affect Its Level

Now the Least Term and Greatest Term total a range for the basic impulse variable. In this case, the (myopic) range is infinite. The $-\infty$ is due to the negative rate of substitution of the transportation activity TMOSELA, which has no upper bound. The $+\infty$ is due to the positive rate of substitution of TMOSFCH, also unbounded.

This is sometimes clearer by expressing the same information in its equation form. Figure 5-30 illustrates this for the same basic impulse. It algebraically describes the basic equation containing activity TMOSEDE as a function of the nonbasic variables.

```
ANALYZE ... RAT  //E
  Rates of substitution for COL TMOSEDE
  ...Status=B Level=20.000  Price=0
             +        1.0   x DMODE
             +        1.0   x DMOLA
             +        1.0   x DMOSE
             +        1.0   x SMOSF
             +        1.0   x DMODE1
             +        1.0   x DMOLA1
             +        1.0   x DMOSE1
             -        1.0   x SMOSF1
             -        1.0   x TMOSELA
             +        1.0   x TMOSFCH
 ------------------------------------------------
       TMOSEDE  Range:         - *           *
```

Figure 5-30. The Equation Format for the Basic Impulse

Another option is the syntax, rather than row and column names. If you have read in the syntax file for WOODNET (or if you EXECUTEd WOODNET to begin this exercise), try specifying RATEOF //SYNTAX. Another option is to restrict the response variables to be only columns or only rows. We shall revisit this in chapter 11, which describes rule files that provide automatic interpretation. Use on-line help and try different examples, say with WOODNET, to experiment with the different output options.

When to use the RATEOF command is a matter of analysis style. It offers a quantitative view of what must change, but the table could become very large. By contrast, using TRACE and PICTURE offer qualitative views from the original matrix. Different problems require different views.

5.3 Using the basis.

Besides its role in computing the rates of substitution, the basis offers some information that may be useful in an analysis problem. The very structure of the basis, illustrated in figure 5-31, may sometimes be regarded as a hierarchy of activities. Such is the case for a network model because the basis can always be rearranged to a lower triangular form, in which case the *kernel* is empty.

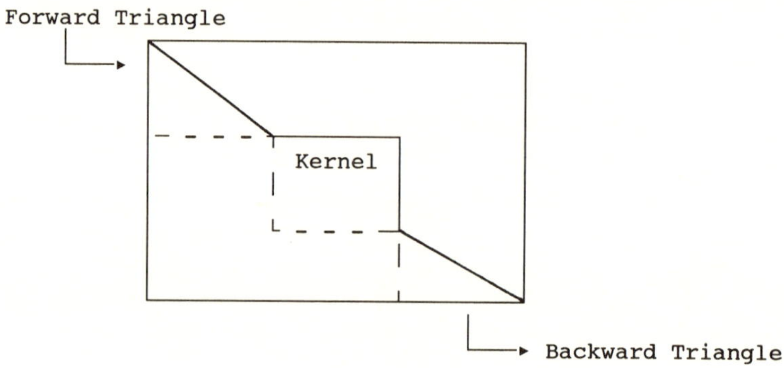

Figure 5-31. Basis Rearrangement Structure

As we proceed down the forward triangle to solve for the basic levels, each value depends on only the preceding ones. In this sense, the associated activities are hierarchically ordered. If the entire basis can be rearranged to a triangular form - that is, if the kernel is empty, the hierarchy is complete and the direction of flow is from Northwest to Southeast.

The dual prices have the opposite direction. The price of the last row is determined first, then the computations proceed from the Southeast to the Northwest. If the kernel is empty, the price hierarchy is the reverse of the quantity hierarchy. In the WOODNET model, for example, consider the delivery path: Supply—▸Transport—▸Demand. The Demand activity is nonbasic at its lower bound. (Except in a degenerate situation, or when revenues are put in the COST row of a demand activity, the Demand activity will have Stat=L.) We thus have a portion of the basis as the 2×2 shown in figure 5-32.

	Transportation Activity	Supply Activity
Demand balance	1	
Supply balance	-1	1

Figure 5-32. A Portion of the Basis Associated with a Non-degenerate Delivery Path in a Network Model

The ordering of the rows and columns is done to obtain the lower triangular form shown. In terms of computing quantities of flow, the transportation activity is computed first, driven by the demand it fulfills. Then, the supply is computed to balance the transportation. Thus, the hierarchy that corresponds to computing quantities of flow is Demand—▸Transportation—▸Supply. In terms of computing prices, the supply price is computed first, which in the non-degenerate case is simply its COST coefficient. Then, the transportation cost is added to determine the Demand price. Thus, the hierarchy that corresponds to price computations is the reverse. This is consistent with the economic concept of exchange, fundamental in activity analysis, where goods and services are exchanged.

For non-network models, the kernel is typically non-empty. Its distinguishing property is that it contains column *spikes* – that is, columns that protrude above the main diagonal. Sometimes, a spike represents competition for resources. Imagine, for example, 3 rows: a resource constraint and 2 product demand constraints. Suppose all 3 of the 3 basic activities require the resource, so the resource row is completely dense. In that case, every basis rearrangement will have at least 1 spike due to this competition for the resource (c.f., §9.3).

The basis structure shown in figure 5-33 may be obtained by specifying `BASIS STATS` (or simply `BA`). The *fill-in* is the number of nonzeroes that are created to solve the basic equations. It is always zero when the basis triangularizes, as in the case of any network model.

```
ANALYZE ... BA
The basis triangularizes
                .
                |\
                | \
                |  \
                |   \
                |____\

...The total number of (stored) nonzeroes =  99...   0 fill-in.
```

Figure 5-33. Basis Statistics for WOODNET

Another use of the basis is to test for *redundant constraints*. A constraint is redundant if its removal leaves the feasible region unchanged. In ANALYZE canonical form, the constraints are the bounds on rows and columns. Thus, the i-th row is redundant if $L_i \le y_i \le U_i$ in every *otherwise feasible solution* – that is, that satisfies all equations and the other bounds; similarly, the j-th column is redundant if $L_j \le x_j \le U_j$ in every otherwise feasible solution. To avoid false conclusions due to computational error, all redundancy detection is that of *strong redundancy* – that is, where a bound holds strictly in every (otherwise) feasible solution.

Some redundancies are trivially detected by computing the range of a row variable, say y_i, just from the bounds of the activities having nonzero coefficients in that row. Then, this range is compared with the given bounds to see if either of them are redundant. Deferring further algorithmic details until chapter 9, let us accept this.

The general problem of finding redundancy in an instance of a linear program is computationally intractable for the general case, but there are simple tests to see if a constraint is redundant, relative to the basis. If no solution file was read into memory, the basis (after simply reading the matrix) is the identity matrix, so all tests are applied to the original coefficients. (We shall see another way to detect redundancies from the original matrix in the next section.)

The redundancies are put into blocks, allowing only one of the bounds to be redundant while the other one may be violated in some, otherwise feasible, solution. The row and column blocks, created by the BASIS REDUND command (abbreviated BA R), are:

RED.LO............Lower bounds are redundant;
RED.UP............Upper bounds are redundant;
RED.FR............Both bounds are redundant.

The associated row and column blocks are created (an error message is given if they already exist). After the tests are complete, empty blocks are erased. Free rows are ignored, so at least one bound is finite for each redundancy. The variables (row and column levels) that have only one (finite) bound are totally redundant if the one bound is redundant - that is, they are put into block RED.FR.

In figure 5-34 we illustrate the redundancy test for the optimal instance of WOODNET, using the optimal basis. We see that 6 activity bounds are totally redundant - that is, both lower and upper bounds are redundant. The other two blocks, RED.LO and RED.UP, were not generated because there were no redundancies for just lower bounds or just upper bounds, respectively.

```
ANALYZE ... SUB *
ANALYZE ... BA R
                                REDUNDANCY TEST
          6 IN COLUMN BLOCK RED.FR
ANALYZE ... BL C D
 1         COL Block
 Block    Mask    Number in Block
 -------------------------------
 RED.FR   ***S     6
```

Figure 5-34. Redundancy Detected With the Optimal Basis

To see the columns in this block, we first clear the submatrix. As shown in figure 5-35, this is done with the COUNT command (abbreviated C) using the row name mask that consists of just the period (.). Because we know there are no rows that match this name mask, the submatrix will be cleared. (The reason for doing it this way is to avoid the SUBMAT sub-command level. For purely interactive processing, this merely saves a keystroke, but when using execution and rule files (see chapter 6), this avoidance of entering the SUBMAT command level is more important.)

Then, we copy from the block to the submatrix. The full specification is BLOCK COLUMN COPY FROM RED.FR, but we abbreviated the command and keywords as BL C C F. The subsequent DISPLAY COLUMN gives the bounds. Since there are no upper bounds, the meaning of redundancy for these 6 columns is that they can never be negative in any feasible solution. Why this is so is our next problem.

```
ANALYZE ... C .
 NO ROW MATCHES .
  Submatrix has  0 rows, 0 columns, and 0 nonzeroes
ANALYZE ... BL C C F RED.FR
ANALYZE ... D C
6 Columns are in submatrix.
COL      STAT  LEVEL (X)     LO_BOUND      UP_BOUND      PRICE (D)
----------------------------------------------------------------
SMOSE1    B        75.000        0             *             0
SOKSE1    B        15.000        0             *             0
SWASE1    B        40.000        0             *             0
TMOSESF   B        30.000        0             *             0
TMOSFSE   B        10.000        0             *             0
TWASFSE   B         5.000        0             *             0
```

Figure 5-35. Displaying the Redundant Columns

In figure 5-36, we picture these columns to begin to understand the redundancies - that is, why these activities must have strictly positive levels in *every* feasible solution. Note that PI * specifies that we want all rows with nonzeroes in the columns of the submatrix; if the * is omitted, the picture will consider the columns as null, because the submatrix currently has no rows.

```
ANALYZE ... PI *
                                   Picture of submatrix
          S S S T T T
          M O W M M W
          O K A O O A
          S S S S S S
          E E E E F F
          1 1 1 S S S
                F E E
 COST     + + + + + + = MIN
 DMOSE            +   =  0
 DMOSF          +     =  0
 DWASE              + =  0
 SMOSE    +     -     =  0
 SMOSF            -   =  0
 SOKSE      +         =  0
 SWASE        +       =  0
 SWASF              - =  0
```

Figure 5-36. Picturing the Submatrix with the Redundant Columns

To understand the redundancies, we want to see these rows in full, except the COST row. By using the SUBMAT command we can alternate between putting in all rows for the submatrix columns (as with the PI * in figure 5-36), then all columns for the result. That is, consider the following sequence.

`PI *` ....picture all rows with the columns in the submatrix
`SUB D R COST` ....delete the cost row
`SUB COL *` ....add all columns with nonzeroes in the submatrix rows (including the ones just added)
`PI` ....picture the (expanded) submatrix

If you keep entering these 3 commands cyclically, you are alternating between bringing in the rows to complete the current columns, then the columns to complete the rows. The intermediate step of deleting the COST row is to avoid bringing in columns (with `SUB COL *`) that merely have a nonzero COST coefficient, whereas we want the columns relevant to satisfying the equations. More generally, this type of alternation may specify `SUB D R * L=-* U=*` to delete all free rows. In our case, COST is the only free row. Another variation is to specify:

`SUB D R * Y GT L Y LT U`

to delete all slack and surplus rows as well as the objective.

A Digression

Whenever you plan repeatedly to enter the same sequence, such as the above, you could prepare an execution file. The next best way to save yourself keystrokes is to set string variables. In our case, we could specify:

```
...STR P PIC *
...STR DC SUB D R COST
...STR AC SUB A C *
```

Then, the same sequence could be specified as:

```
...'P
...'DC
...'AC
...PI
```

The single quote (') denotes a string, where ANALYZE substitutes `PIC *` for `'P`, etc. You could also specify `SW ECHO` to have the commands echoed, which is done after string substitution. Now let us resume our analysis.

Observe, in figure 5-37, that demand rows DMOSE, DMOSF and DWASE have only one positive coefficient to balance the negative coefficient of the associated demand activity. This fixes the levels of TMOSFSE, TMOSESF and TWASFSE, respectively. This basis revealed those redundancies - that is, their levels could never be negative, which renders their lower bounds of 0 redundant. This explains why the associated supply activities (SMOSE1 and SWASE1) must have positive levels in every feasible solution. It does not, however, explain why the level of SOKSE1 must be (strictly) positive.

```
ANALYZE ... SUB D R COST
ANALYZE ... SUB COL *
ANALYZE ... PIC
                    Picture of submatrix
        D D D S S S S S T T T T T T T T T T T T T T T T T T T T
        M M W M M O W W M M M M M M M M O O O O W W W W W W W W
        O O A O O K A A O O O O O O O O K K K K A A A A A A A A
        S S S S S S S S S S S S S S S S S S S S S S S S S S S S
        E F E E F E E F E E E E F F F F E E E E E E E E F F F F
        1 1 1 1 1 1 1 1 C D L S C D L S C D L S C D L S C D L S
                        H E A F H E A E H E A F H E A F H E A E
DMOSE   -                             +                         = 0
DMOSF     -                   +                                 = 0
DWASE       -                                                 + = 0
SMOSE         +         - - - -                                 = 0
SMOSF           +               - - - -                         = 0
SOKSE             +                     - - - -                 = 0
SWASE               +                           - - - -         = 0
SWASF                 +                                 - - - - = 0
```

Figure 5-37. Completing the Columns.

To understand the (strong) redundancy of the 0 lower bound on the supply activity SOKSE1, we first trace through some pictures, in order to illustrate how to build successive submatrices that eventually help us to put the story together. Afterwards, we shall show an easier way.

Figure 5-38 shows the beginning of our navigation by first picturing the supply activity (SOKSE1), then showing its rim and matrix values with the SHOW COLUMN command (abbreviated SHO C).

```
ANALYZE ... PI *,SOKSE1
                         Picture of submatrix
        S
        O
        K
        S
        E
        1
COST    + = MIN
SOKSE   + =  0
ANALYZE ... SHO C
 COL SOKSE1   HAS 2 NONZEROES
 BOUNDS:             L=0            U=*
 SOLUTION:  STAT=B  X=15.000        D=0
 Row      Coefficient     Price       Factor
----------------------------------------------
SOKSE           1.0          50.000    -  50.000
COST          50.000       - 1.0          50.000
==============================================
                              SUM:        0
```

Figure 5-38. Beginning the Navigation to Understand the Redundancy of the Oak Supply Activity (SOKSE)

To proceed with the navigation, we want to see the activities adjacent to the supply row (SOKSE) - that is, columns having a nonzero coefficient in the row. To do so, we first delete the COST row from the submatrix, as shown in figure 5-39, then we put all columns into the submatrix. The subsequent picture retains only those columns having a nonzero in the (only) submatrix row. This tells us that the Seattle (SE) supplier of oak (OK) is linked to Chicago (CH), Denver (DE), Los Angeles (LA), and San Francisco (SF). Our next step is to find out who else is linked to these consumers.

```
ANALYZE ... SUB D R C
ANALYZE ... SUB CO *
ANALYZE ... PI
                                Picture of submatrix
         S T T T T
         0 0 0 0 0
         K K K K K
         S S S S S
         E E E E E
         1 C D L S
           H E A F
SOKSE    + - - - - = 0
ANALYZE ... PI *
                                Picture of submatrix
         S T T T T
         0 0 0 0 0
         K K K K K
         S S S S S
         E E E E E
         1 C D L S
           H E A F
COST     + + + + + = MIN
DOKCH      +       = 0
DOKDE        +     = 0
DOKLA          +   = 0
DOKSF            + = 0
SOKSE    + - - - - = 0
```

Figure 5-39. Completing the Columns, Then the Rows, to Picture the Supplier's Links for Satisfying Demands

Figure 5-40 continues by picturing all rows that are adjacent to the columns in the submatrix. Then, figure 5-40 deletes the COST row before putting all columns into the submatrix. This continues the navigation by picturing all activities that are linked to the consumers linked to our original supplier – that is, whose redundancy we seek to explain.

```
ANALYZE ... SUB D R C
ANALYZE ... SUB CO *
ANALYZE ... PI
                       Picture of submatrix
        D D D D S T T T T T T T T T T T T T T T T
        0 0 0 0 0 0 0 0 0 0 0 0 0 0 0 0 0 0 0 0 0
        K K K K K K K K K K K K K K K K K K K K K
        C D L S S C C C D D D L L L S S S S S S S
        H E A F E H H H E E E A A A E E E E F F F
        1 1 1 1 1 D L S C L S C D S C D L S C D L
                  E A F H A F H E F H E A F H E A
DOKCH   -               +     +     +       +     = 0
DOKDE     -       +             +     +       +   = 0
DOKLA       -       +     +             +       + = 0
DOKSF         -       +     +     +       +       = 0
SOKSE           +                   - - - -       = 0
```

Figure 5-40. Picturing the Other Activities Linked to the Supplier's Distribution Network

We may continue in this fashion, displaying column bounds along the way to see why the Seattle oak supplier must have a positive level in every feasible solution. The result is depicted graphically in figure 5-41.

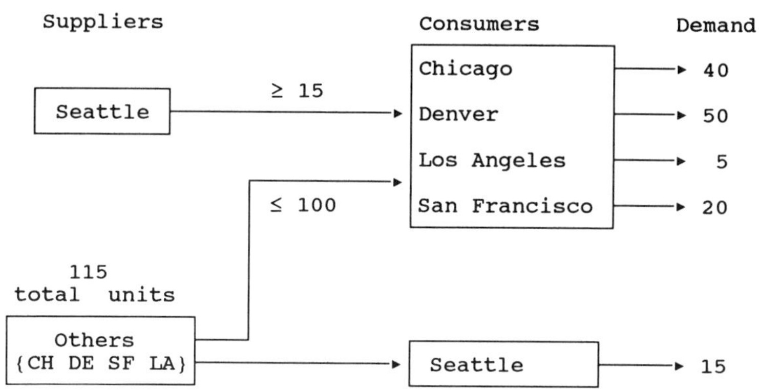

Figure 5-41. Oak Distribution Network to Explain Why the Level of the Seattle Supplier is Always Positive

Now consider an easier way to obtain this information without navigating through the matrix as we have done. Figure 5-42 shows the rates of substitution for this supply activity. It shows the basic equation that describes why the level of SOKSE1 must be positive in every feasible solution: to meet the demands. Let us elaborate upon (and verify) what this says.

As shown in figure 5-42, the basic equations reveal that the range of this level is [15, 130], which renders its lower bound (0) redundant. In general, the RATEOF command can be used as one way to explore the redundancies, especially if they not are detected from the original matrix. (You can shorten the output by entering //COL as an option, following the RATEOF specification.)

```
ANALYZE ... RAT C SOKSE1
  Rates of substitution for COL SOKSE1
  ...Status=B Level=15.000  Price=0
 Nonbasic    Rate of        Least          Greatest
 Variable  Substitution    Term           Term
 =================================================
 DOKCH             1.0          0              0
 DOKDE             1.0          0              0
 DOKLA             1.0          0              0
 DOKSE             1.0          0              0
 DOKSF             1.0          0              0
 SOKCH             1.0          0              0
 SOKDE             1.0          0              0
 SOKLA             1.0          0              0
 SOKSE             1.0          0              0
 SOKSF             1.0          0              0
 DOKCH1            1.0          40.000         40.000
 DOKDE1            1.0          50.000         50.000
 DOKLA1            1.0           5.000          5.000
 DOKSE1            1.0          15.000         15.000
 DOKSF1            1.0          20.000         20.000
 SOKCH1          - 1.0      -   40.000         0
 SOKDE1          - 1.0      -   50.000         0
 SOKLA1          - 1.0      -    5.000         0
 SOKSF1          - 1.0      -   20.000         0
 -------------------------------------------------
       SOKSE1    Range:         15.000        130.000
```

Figure 5-42. Displaying the Rates of Substitution for a Redundant Supply Activity

Because all rows are homogeneous equations, the nonbasic row variables that affect the supply activity (SOKSE1) do not contribute to its range. The demands, however, linked to this supplier have unit rates of substitution – that is, any change in demand from Chicago, Denver, Los Angeles or San Francisco changes the level of the Seattle supplier by the same amount. Then, we see negative rates of substitution with those supply activities that are linked to these same consumers. Their contribution to the range of SOKSE1 comes from their bounds. The net result is that the Seattle supplier must have a level between 15 and 130. The least value (15) can be explained from the RATEOF display, which can be organized (manually) to give the graphical depiction in figure 5-41.

If this supplier had an upper bound of more than 130, it would also be redundant. This is because its range is limited by 130 due to the fixed demands that sum to 130. In fact, however, it cannot supply more than 115 because it is not linked to the Seattle consumer. The unit rate of substitution with the demand activity DOKSE1 causes an overestimate of the maximum level in its range. In this case, the rate reflects the fact that if the demand in Seattle were to increase, this would require units from the other suppliers, thus reducing the amount they can send to Seattle's customers.

To further illustrate the dependence on the basis in the resident solution, we re-read the WOODNET matrix without the optimal basis, as shown in figure 5-43. Then, we use the BASIS command (abbreviated BA) just to setup the basis. After putting all rows and columns in the submatrix, we ask for redundancy detection. Here the basis is the identity matrix, and no redundancy is detected.

```
ANALYZE ... REA M WOODNET
NAME     WOODNET
         36 rows...              1 free
         93 columns...            221 nonzeroes
 RHS section complete...NAME = HOMOGEN
 BOUNDS complete...NAME = BNDSET
 Matrix read in
 Initialized solution to all logical basis.
ANALYZE ... BASIS
 Basis must be setup...please wait
 Submatrix will be cleared...please reset after basis is setup
 ...To avoid this next time WRITE new packed file after basis is setup
ANALYZE ... SUB *
ANALYZE ... BAS R
                                REDUNDANCY TEST
 ...NO REDUNDANCY FOUND
```

Figure 5-43. No Redundancy is Detected for the Same Instance of WOODNET, but for the Logical Basis

5.4 The REDUCE command.

The REDUCE command iteratively seeks reductions in bounds for the levels of primal quantities and dual prices. Sometimes reductions beget other reductions to the point where fixed levels are detected. Figure 5-44 shows the execution for the optimal instance of WOODNET. After reading the WOODNET packed file into memory, the OUTPUT command (abbreviated OUT) spools the output to the file named WOODNET.RED, which you should peruse later. In general, the output from REDUCE is extensive because it logs every reduction, and there is no scrolling. If you do not specify an OUTPUT file, the reductions will flash across the screen too fast to read them, so it is good practice to spool the output. You can use the PRINT command afterwards to see the output file, but it is even better if you have a windows environment to use an editor in another window.

```
ANALYZE ... REA P WOODNET
 Packed LP read from WOODNET.PCK

ANALYZE ... OUT WOODNET.RED
ANALYZE ... REDUCE *,*
REDUCE PASS    1
REDUCE PASS    2
REDUCE PASS    3
REDUCE PASS    4
REDUCE PASS    5
REDUCE PASS    6

        Summary of Reductions
        ----------------------
   Rows:     0 freed    3 forcing    0 reduced    0 weakened
Columns:     3 fixed    0 forcing   58 reduced

ANALYZE ... OUT *
```

Figure 5-44. Executing the REDUCE Command for the WOODNET Matrix

The *Summary of Reductions* shows 3 forcing rows. This means three of the rows forced the levels of their adjacent activities. This agrees with the column summary, which shows 3 columns are fixed. This means that 3 activity levels that were not already fixed a'priori (as the demand activities were) were inferred to have only one value in every feasible solution. There are also 58 columns whose bounds are reduced from their original values. The last specification, 0 *, restores output to come to the screen. If you examine the output file, WOODNET.RED, you will see the reasons for the reductions and a final table of activities whose bounds are reduced.

The REDUCE command can also be used to help diagnose an infeasibility. Figure 5-45 illustrates this with the infeasible instance of WOODNET. After reading WOODINFE into memory, REDUCE is executed. After 1 pass, the infeasibility is detected. (The summary of reductions is not relevant, but you will see it on your screen.) At the ANALYZE prompt, enter 0 * to return the output spooling to the screen. You may now picture this row to verify the result, which we analyzed in chapter 4.

```
ANALYZE ... REA P WOODINFE
 Packed LP read from WOODINFE.PCK

ANALYZE ... OUT WOODINFE.RED
ANALYZE ... REDUCE *,*
REDUCE PASS     1
!Primal infeasibility detected with row DMOSE
Required value range:    .00000000     to   .00000000
Computed value range:  -.10000000E+21 to  -10.000000
```

Figure 5-45. Executing the REDUCE Command for an Infeasible Instance of WOODNET

If you examine the output file, WOODINFE.RED, you will see the summary information shown in figure 5-46. For the row where infeasibility is detected, the table lists its activities (i.e., those with nonzero coefficients in the row) plus the relevant bound that contributed to the extreme value in the row's computed value range. In this case, the row has only one activity, namely DMOSE1, and its coefficient is –1. The relevant bound is that the level of DMOSE1 ≥ 10 (= L), which makes the maximum value of row DMOSE = –10. Try this for other examples, like BLENDINF (see MOREDATA.DOC and Appendix B).

```
                             Relevant        Value of Term
 Activity      Coefficient   Bound           (Coef * Bound)
 DMOSE1           -1.0000    L=  10.000          -10.000
                                               SUM =  -10.000
```

Figure 5-46. Last Portion of WOODINFE.RED (c.f., Figure 5-45)

The REDUCE command can sometimes detect unboundedness and give a good starting point for diagnostic analysis. You may try this for the unbounded instance of WOODNET by reading WOODUNBD into memory. You will find, however, that REDUCE fails to detect this one.

The successive bound reduction algorithm, executed by the REDUCE command, can also recognize integer variables to strengthen the reductions. binary variables have additional properties when some constraints represent logical expressions, such as: IF u=1 THEN v=1, which is represented by the inequality, u–v ≤ 0. Figure 5-47 illustrates this with a small test problem called LGLTEST.MAT.

```
ANALYZE ... REA M LGLTEST
NAME    LGLTEST
            5 rows...            1 free
            4 columns...           10 nonzeroes
 RHS section complete...NAME = RHS
BOUNDS complete...NAME = BNDSET
 Matrix read in
 All rows are basic
 Initialized solution is OPTIMAL
ANALYZE ... L *,*
    0  >= R1     =  X1 - X2
    0  >= R2     =  X2 - X3
    0  >= R3     =- X1 + X3
  1    >= R4     =  X1 + X2 + X3 +  10 I4
ANALYZE ... O LGLTEST.RED
ANALYZE ... RED  //L, B=X, I=I
 REDUCE PASS    1
 REDUCE PASS    2

          Summary of Reductions
          ----------------------
   Rows:      0 freed     0 forcing     0 reduced     0 weakened
Columns:      1 fixed     0 forcing     0 reduced

         6 Implications Written
ANALYZE ... O *
```

Figure 5-47. Executing REDUCE with Logical Testing

After reading the matrix into memory, we list its equations. The three X-variables are binary, and I4 is integer-valued. The REDUCE command is specified with the parameter specifications to identify the binary and integer variables by name masks (B=X and I=I, respectively); and, the LOGICAL parameter (abbreviated L) specifies that the implications test is to be performed.

If you look at the output file, LGLTEST.RED, you will see the implications found, as shown in figure 5-48. The implications reveal that all variables must be 0 in every feasible solution.

```
                     BEGIN LOGICAL IMPLICATIONS TEST
** FROM ROW R1
 IF X1        = 1 THEN X2       = 1
** FROM ROW R2
 IF X2        = 1 THEN X3       = 1
** FROM ROW R3
 IF X1        = 0 THEN X3       = 0
** FROM ROW R4
 IF X1        = 1 THEN X2       = 0
 IF X1        = 1 THEN X3       = 0
** FROM ROW R4
 IF X2        = 1 THEN X3       = 0
```

Figure 5-48. The Logical Implications Written to the Output File (LGLTEST.RED)

In the interests of model management, it helps to understand *why* reductions occur. Presently, this is not fully automated, but the successive bound reduction algorithm (described in §9.6) writes a message for each reduction to the output file. You may chain through this log, preferably with an editor in a windows environment, so you can also use ANALYZE in another window.

5.5 The AGGREGAT command.

The AGGREGAT command displays the aggregation of rows in the submatrix (using all columns in the entire matrix). This is a linear combination of the rows, weighted by the resident dual prices. Its primary purpose is to see a 1-constraint infeasible implication when the resident solution is a true phase I solution.

The underlying theory is an early result by Dantzig:

Theorem. Suppose $\{(x,y)\colon y=Ax,\ L \le x \le U,\ \lambda \le y \le \mu\}$ is empty, and suppose π is a phase 1 dual price (m-vector). Then, the following system is infeasible: $\pi^+\lambda + \pi^-\mu \le \pi Ax \le \pi^+\mu + \pi^-\lambda,\ L \le x \le U$.

One can apply the AGGREGAT command to the full matrix to obtain the coefficients, πA, and the aggregate range. An arbitrary submatrix can be specified to see a subsystem that is infeasible. (If the subsystem is not trivially infeasible – that is, whose aggregate range is inconsistent – it is inconsistent with the rest of the rows, not in the submatrix.)

Figure 5-49 illustrates with the infeasible variation of WOODNET, called WOODINFE (c.f., figure 5-45). After reading in the packed file, the submatrix is set to the entire LP matrix. The aggregate constraint, which is displayed, shows the implied infeasibility:

Aggregate equation: – DMOSE1 – DWASE1 = 0

where = 0 is the Aggregate ROW range. The Myopic range, computed from the bounds on the two activities, shows that the left-side of the equation cannot exceed -15, thereby making the equation infeasible.

```
ANALYZE ... REA P WOODINFE
 Packed LP read from WOODINFE.PCK

ANALYZE ... SUB *
ANALYZE ... AGGR R
  Aggregate ROW range:        0     to        0
  Coefficient:                        Range
       - 1.0      DMOSE1            -*                    -  10.000
       - 1.0      DWASE1            -*                    -   5.000
            ...Myopic range:      - *      to    -  15.000
```

Figure 5-49. Using the AGGREGAT Command

Sometimes this display can reveal what went wrong with the model specification or the data. There is, however, an anomaly, which is described in chapter 7.

Chapter 6. File Interfaces

A parity error is a bit funny.
James E. Kalan

In this chapter we describe the types of files, which may be input or output, for ANALYZE. Some of these are for environment control, so we start with an explanation of special files that affect the start-up process. Then, we describe how you may enhance the on-line help, especially for others to use, with dictionaries and documents. In addition, you may prepare a sequence of ANALYZE commands that comprise an *execution file*. Examples include some that were executed in chapters 4 and 5.

The only required input file to ANALYZE is a *matrix file*. This is the de facto standard that was designed in the early years of linear programming (1950's). This may be read into memory with the READIN command, or it may have undergone some revision by using the LPRENAME utility (see chapter 8). In our exercises of the previous chapters, the matrix files were read in, followed by a solution file input. The resulting information was written to a packed file for quick loading. These input procedures were hidden during the exercises, having done them in ATEST.EXC. Now, to elaborate, the matrix and packed files are described in §6.4.

Although ANALYZE can be useful without a solution of the particular instance, it is typical that a solution file is generated by some optimizer, and the purpose of using ANALYZE is either to debug a problem, such as infeasibility, or to probe into the meaning of the solution. In §6.5, the solution files recognized by ANALYZE are described.

The English translations for the EXPLAIN command are determined by the model's syntax, which is described by a *syntax file*. A syntax file is generally associated with the entire model, rather than with just one instance. For example, if you examine WOODNET.EXC, WOODINFE.EXC and WOODUNBD.EXC, you will find that each execution file reads the same syntax file into memory, namely, WOODNET.SYN. In §6.6 the basic structure of a syntax file is described.

A *rule file* provides a means to give intelligent assistance for analysis. It supports the INTERPRT command, which was illustrated in §4.2. This is explained in §6.7.

Sometimes it is desirable to save a submatrix, a set of blocks, or a schema, for later loading. These are described in §6.8.

Most files described in this chapter are ASCII, and some are limited to record lengths of 80 bytes. All ASCII file inputs to ANALYZE can have an asterisk (*) in column 1 for a comment, which is ignored. Blank lines are also ignored.

Most of the ASCII input files recognize an INCLUDE statement (an exception is the matrix file). In an execution file, an INCLUDE statement allows execution of another file of commands and continuation with the main one (but the RETURN statement still returns to the terminal). In syntax and rule files this allows some decomposition of their contents. The only restrictions are:

1. The keyword INCLUDE must begin in column 1, followed by the INCLUDE file specification.
2. Only one INCLUDE file can be active - that is, you cannot have an INCLUDE specification within another one.

The second restriction necessitates caution. For example, if you use an INCLUDE statement in an execution file that reads a syntax file, the syntax file cannot contain another INCLUDE statement. In addition, you must not specify READIN IIS from an INCLUDE file, because the input will be mistakenly read from the INCLUDE file. The best way to become familiar with the use of the INCLUDE statement in an ASCII file input is to examine ATEST.EXC, WOODNET.SYN and TEST.RUL for examples.

The last section suggests some ways to setup your environment for analysis by using the various file types in different ways. It is recommended that you setup your files to suit your environment. At the very least, examine carefully the use of execution files not only to setup the environment, but also to read LP files into memory with syntax files and with an introductory message to your constituent. An example of this is in WOODNET.EXC, which we used in chapter 4.

On-line help is available with FILES.DOC (use the PRINT command).

6.1 Environment controls.

There is a variety of *session control commands* that help you control the environment, and you may obtain on-line help with each of these. Here we shall describe some that are relevant to files and their use.

Each file specification consists of a prefix, a filename, and a suffix. The _SETUP command allows you to change the prefix and suffix of each type. You may enter _SETUP to see the current defaults within ANALYZE. After installation, all prefixes are null, which means your default path is used. Each suffix (in the DOS version) corresponds to the type of file, and it too may be displayed or changed.

Before the session begins, there are special files sought by the user interface module of ANALYZE (called FLIP). Figure 6-1 lists these and the functions they perform. If the special file is absent, no action is taken and no message is printed.

Name	Type	Action
SETUP	EXECUTION	Change prefixes and suffixes
ANALYZE	DICTIONRY	Load dictionary
CREDITS	PRINT	Print credits
n/a	KEYWORD	Load keywords
PROFILE	EXECUTION	Execute commands
RULBASE	RULE	Load rules, which are options for the INTERPRT command

Figure 6-1. Special Files Sought Before the Interactive Session Begins

The special files are listed in order of interrogation. For example, the SETUP.EXC file may be used to set the prefix or suffix of execution files, in which case the PROFILE sought is with the new prefix and suffix. Upon installation, the PRINT suffix is .DOC, so the file CREDITS.DOC is printed. There are no KEYWORD files for ANALYZE, but you may add this. See the on-line help for how it can be used (also see RANDMOD for the use of keyword files to generate spec files for OB1 and MINOS).

Use the PROFILE to set screen parameters (with the SCREEN command), to set strings that may be useful for reducing the number of keystrokes (with the STRING command), to set file defaults (with the _SETUP command), to load or merge dictionaries (with the _DICTNRY command), and to print an introductory message about the LP instance.

After the start-up process, the _SETUP command allows you to change prefixes and suffixes of recognized file types during the session. Figure 6-2 shows the result of entering _SETUP to see the current defaults (for DOS environment).

```
ANALYZE... _SETUP
                          _SETUP Entered
      Keyword        Prefix                           Suffix
      ------------------------------------------------------
      DICTNARY                                        .DIC
      EXECUTE                                         .EXC
      HELP                                            .HLP
      INCLUDE                                         .INC
      KEYWORD                                         .KEY
      PRINT                                           .DOC
      BLOCK                                           .BLK
      FORMAT                                          .FPK
      GRAPH                                           .GPH
      IIS                                             .IIS
      MATRIX                                          .MAT
      MINOS                                           .MNS
      OB1                                             .OB1
      OSL                                             .OSL
      PACKED                                          .PCK
      RULEBASE                                        .RUL
      SCHEMA                                          .SCH
      SUBMAT                                          .SUB
      SYNTAX                                          .SYN
```

Figure 6-2. Seeing the Default Prefixes and Suffixes of Filespecs with the _SETUP Command

Suppose, for example, you specify READIN MATRIX WOODNET. The full filespec is WOODNET.MAT with the above defaults. But suppose you want to read in a matrix file whose full filespec is A:\MYPROB\WOODNET.MAT. Then, you may either fully qualify it as:

READIN MATRIX A:\MYPROB\WOODNET.MAT

or you may enter _SETUP MATRIX A:\MYPROB\ to change the default prefix and specify:

READIN MATRIX WOODNET

In the following sections, the various file types are explained.

6.2 Dictionaries, documents, help, and keyword files.

A dictionary is a file with entries of the form:

key := *text*

where *key* is any 1 to 16 character expression, and *text* is what is printed when the user asks ? *key*. The delimiter := separates the *key* from the *text*. The _SETUP file type is DICTNARY.

Examine the file ANALYZE.DIC, which is the dictionary that came with your ANALYZE system. The *text* may continue to the next line; it ends with another *key* definition on a new line or when the end of the file is reached.

To create your own dictionary you simply create an ASCII file with 80-character record lengths and enter the key and text, separated by := as described above. ANALYZE does not sort the keys, so you must do that with your editor. ANALYZE simply treats this as a list and searches linearly for key matches. This suggests dictionaries should not be too big; they are intended to provide quick reference to a small number of terms.

A document, or PRINT file, is also created with your own editor and is presumed to be 80-character record lengths. ANALYZE truncates each line to fit in the screen width set by the user, so it is advisable to have left and right margins. The file is printed sequentially, so they should be short, such as those that came with ANALYZE.

The HELP files are those that came with ANALYZE to give on-line help for each command. Generally, you will not want to change this with _SETUP during a session. If the help files are not located in the default path, you should specify this change in either SETUP.EXC or PROFILE.EXC.

The KEYWORD file type is used to support the _KEYWORD command. There are no keyword files that came with ANALYZE, so you can ignore this (it is used in special environments, such as to set spec files for OB1 or MINOS).

6.3 Execution files.

It is sometimes desirable to prepare a sequence of ANALYZE commands and simply enter EXECUTE *filespec* at the ANALYZE prompt. The *filespec* is the EXECUTE file type recognized by the _SETUP command. The default upon installation under DOS is no prefix (i.e., your default path on your hard disk), and the suffix is .EXC (the dot is needed). You may examine the EXECUTE files that came with ANALYZE by entering DIR *.EXC at the DOS prompt. In particular, we executed WOODNET, WOODINFE, and WOODUNBD in chapters 4 and 5.

To control message spooling, there are several ways to print a comment to the screen and/or output file. In general, any ASCII input file (including the terminal) can have an asterisk in column 1 to indicate a comment. This is completely ignored; however, a dollar sign ($) in column 1 of an execution file echoes the comment to the screen. The comment can also be directed to the output file, if not the screen, by using two dollar signs ($$) in columns 1 and 2. An example of this is in DOCANAL.EXC. The screen output is clipped at the screen width, but the filed output is treated like text in that it is fully printed even if more than one line is required.

See, also, the _PAUSE and _TYPE session control commands. The first issues a pause, much like the DOS PAUSE command, allowing the user to abort the execution if something is not set right. The latter types the line to the output file. Unlike the double dollar signs, the _TYPE command does not type to the screen; and, it has justify options (left, right and center). Also look at the switches, such as SWITCH ECHO_ON (or simply SW ECHO), which toggles the echo. If this is on, the commands are echoed to the screen and (if different) to the output file. This is useful for letting the user know the commands that are being executed.

Sometimes it is desirable to suppress messages appearing on the screen, such as when reading a matrix file into memory. To do so, set the message switch off by specifying SWITCH MSG_ON (or simply SW M). All switches toggle, so remember to specify this again before leaving the execution file.

You might see ATEST.EXC, which you executed just after installing ANALYZE. It illustrates most of the session controls you might want to have in your own execution file. One exception is the error switch, ERR_STOP. If you turn this off, the execution continues even if an error is encountered; otherwise, the execution file is aborted, returning to the screen for command input.

Another feature of execution files is *looping*. The syntax is to put an integer after the file specification. To illustrate, consider the following two examples.

Suppose FILE.EXC contains the following commands.

```
READ M 'NAME
SOL OB1 'NAME
BAS
WRITE 'NAME
STRING INCREMENT NAME
```

Then, at the terminal suppose we specify:

```
STRING NAME=CASE0
EXEC FILE 3
```

The execution performs as follows. The first time it reads a matrix file called CASE0.MAT into memory, then its OB1 solution file, named CASE0.OB1. The BASIS command is then executed to setup the basis for later analysis support (e.g., the RATEOF command). Finally, CASE0.PCK is written. The last command increments the string NAME, which becomes CASE1. The second time through the file, therefore, the matrix and solution files for CASE1 (CASE1.MAT and CASE1.OB1, respectively) are read into memory. The basis is setup, and CASE1.PCK is written for later analysis. After completing the second loop through this execution file, NAME is incremented to CASE2.

This example illustrates how one can prepare multiple cases whose file names differ in the last character, like CASE0, CASE1, CASE2, ..., CASE9, CASEA, ..., CASEZ. The STRING INCREMENT allows 36 cases using a single digit or letter for the last character.

Here is a second example that arose in our analysis. Sometimes we want to build a submatrix by alternating between completing the columns for the rows in the submatrix and completing the rows for the columns. A sequence that does this is:

```
SUB COL *
PIC
_PAUSE
PIC *
_PAUSE
SUB D R * L=-* U=*
```

When this sequence of commands is executed, all columns are put into the submatrix, but the picture removes null columns – that is, the rows that were in the submatrix to start are completed. The _PAUSE command allows you to see the picture before continuing with the execution; and, you can abort execution by answering N to the prompt for continuation. The next picture completes the columns by putting all rows into the submatrix, then removing the null ones before picturing. After another pause, free rows are deleted from the submatrix, which includes the objective row.

If this file is executed two times (EXEC FILE 2), the result is to complete the rows, then the columns, then the (enlarged set of) rows, then the (enlarged set of) columns. You might try this with the exercise in §5.3 of explaining the redundancies detected by the BASIS command.

6.4 Matrix and packed files.

A matrix file is the (de facto) standard MPS file, generally recognized by optimizers. Figure 6-3 gives its structure, which is fixed-field. Row and column names are normally limited 8 characters each, but you can specify a name length with the option LEN_NAME=*number*, where *number* is at least 4 and at most 10. If you need more than 10 characters in a name, you cannot read a standard matrix file (use LPRENAME, described in §8.2).

```
NAME                 name
ROWS
 tp      name
COLUMNS
         column      row        nonzero
RHS
         name        row        nonzero
RANGES
         name        row        nonzero
BOUNDS
 tp      name        column     [nonzero]
ENDATA
```

Figure 6-3. The Structure of a Standard (MPS) Matrix File

The ROWS segment has one line per row. In column 2 is the row type (*tp*):

N ....free row (like the objective);
L.................less-than-or-equal-to ($Ax \le b$);
G................greater-than-or-equal-to ($Ax \ge b$);
E ....equation ($Ax = b$).

Then, the row name is specified in column 5.

The COLUMNS segment has one line per nonzero (it could specify two nonzeroes for a column on one line). First, the column name appears in column 5, then the row name in column 15. The nonzero appears in decimal (F) or scientific (E) format in column 25. All nonzeroes of a column must appear sequentially - that is, you may not intermingle nonzero specifications arbitrarily.

The RHS (right-hand side) segment has one line per nonzero right-hand side. First, the name of the RHS set appears in column 5. (In the early period of linear programming, multiple RHS sets were specified at once to represent different scenarios for efficient processing; today, this feature is not generally used, so the RHS name is typically something innocuous.) Then, the row name is given in column 15, and the nonzero is in column 25.

The RANGES segment is used if some L-rows or some G-rows have ranges. That is, an L-row represents the constraint $Ax \le b$. Then, if $r>0$ is a range value, the additional constraint is $Ax \ge b-r$. Similarly, a G-row represents the constraint $Ax \ge b$, so the range specification adds the constraint $Ax \le b+r$.

The range set name is given in column 5 (another carryover from early LP), and the row name is in column 15. The range nonzero value is given in column 25.

The BOUNDS segment is used if bounds are present on the (column) activity levels, or if a level is not restricted to be non-negative. There is one line per bound specification. The type (*tp*) in columns 2-3 is one of:

UP ....upper bound
LO ....lower bound
FX..............fixed level
MI..............lower bound $= -\infty$
PL..............upper bound $= +\infty$
FR..............free (lower bound $= -\infty$ and upper bound $= +\infty$)
BV ....binary variable.

Following the type of bound specification is the name of the *bound set* (another carryover) in column 5. The column name is given in column 15. If the bound type is UP, LO or FX, the nonzero value is in column 25.

The last line of a matrix file is the ENDATA specification. Comments can appear below this, as ANALYZE stops reading the matrix file once it encounters the ENDATA line.

There are special extensions of this basic matrix file, which are supported by ANALYZE. The best place to obtain additional information is from system manuals, like MPSIII. For our purposes, those details are not necessary. You may look at the matrix files that came with ANALYZE, which have the suffix .MAT.

It is sometimes desirable to save a matrix file, after it is read into memory, as a binary file for later analysis. The reason is that the binary file is much quicker to read into memory since all the sorting and other organizations of data structures are done. The save is with the WRITEOUT command, and the subsequent read is with the PACK option of the READIN command. For example, to read WOODNET.PCK into memory, specify:

READIN PACK WOODNET

To change computer environments, the packed file can be formatted. For example, a large model may need to be solved on a mainframe computer. The WRITEOUT command has the option to specify FORMAT, which creates a formatted file version of the packed file, so it can be downloaded onto a workstation. Then, you can specify:

READIN PACK WOODNET FORMAT

(or simply REA P WOODNET F). The _SETUP file type is FORMAT, which is distinct from the PACKED file type.

6.5 Solution files.

Different optimizers usually have different formats in the solution file output. ANALYZE recognizes the following as solution file types for the SOLUTION command:

MINOS..........from Stanford University and Ketron Management Science.
OB1. ....from XMP, Inc.
OSL. ....from IBM.
MPS. ....unformatted, following IEEE standard to represent floating point numbers.
XMP ....same as OB1.

The first three solution file types are formatted (ASCII) files, generated by the respective optimizers. An XMP solution file is the same as the OB1 type. If OB1 is used to solve the linear program, you must specify SIMPLEX YES in the OB1 SPEC file to obtain a basic solution before entering ANALYZE. The MPS solution file type has been tested with MPSX and MPSIII under TSO. It can also be used to read a CPLEX solution file in a SUN 4 environment. It is not available under DOS. See SOLVERS.DOC for further information.

A typical sequence, which can be put into an execution file, is:

READIN MATRIX *name*
SOLUTION *type* [*name*]

This first reads the matrix file, then the solution file. If the name of the solution file is the same as the problem name, it need not be specified. In particular, this is the case for the optimal instance of WOODNET, where WOODNET.MNS is the MINOS solution file. On the other hand, this is also the problem name for the infeasible instance of WOODNET, whereas the MINOS solution file is WOODINFE.MNS, so you must specify:

SOLUTION MINOS WOODINFE

(or simply SOL MI WOODINFE).

If you use an optimizer that is not one of the recognized ones, you can create your own solution file that conforms with one of the recognized ones, then use the SOLUTION command. It is preferable, however, that you then use the RENAME command to rename the solver to the correct one. The SOLUTION command uses the solver type as the description of which optimzer was used, and it appears with the SUMMARY command. It would mislead others if you do not rename it.

The simplest solution type is OSL, which has the format shown in Figure 6-4. Line 1 is a header that first gives the problem name (which is the name given in the matrix file). Then, *opt* is the sense of optimization, which is MIN or MAX, and *status* is the solution status: OPTIMAL, UNBOUNDED, INFEASIBLE, FEASIBLE or UNKNOWN (an INFEASIBLE status means the optimzer determined that the linear program is infeasible, not just that the current solution is infeasible). The right-hand side set, range set and bound set are those given in the matrix file. If the range or bound set is absent, enter `(none)`. The set names and the problem name are used to verify that the solution file agrees with the computer-resident information. A warning message will appear if these do not agree.

```
name  opt  status  rhs_set  range_set bound_set     | Header
name  status  level  price                          | Rows section
...                                                 ...1 line per row
name  status  level  price                          | Columns section
...                                                 ...1 line per column
```

Figure 6-4. OSL Solution File Format

Following the header line, there is one line per row, followed by one line per column. Each line contains the name of the row or column, its solution status (B, I, L or U), its level and its price. An example is in WOODNET.OSL, which came with your ANALYZE data.

It is desirable to prepare such files for an ANALYZE session, then use the WRITEOUT command to save both the matrix and solution as a packed file for a quick readin to begin analysis.

Another type of file is called the IIS file. This is John Chinneck's representation of an *irreducible infeasible subsystem*, which can be useful for diagnosing an infeasible instance. Recall that the canonical form is

$$\text{LP:}\quad \text{opt } cx\text{: } y = Ax,\ L \le (x, y) \le U,$$

where opt can be min or max, and bounds can be infinite (i.e., absent). When LP is infeasible, the objective becomes irrelevant, and at least one bound is violated by the resident solution.

An IIS is a set of equations and bounds such that it is infeasible – that is, there is no solution to the subsystem; and, no equation or bound can be dropped without becoming feasible. The following properties can be shown.

Property 1. An equation is in the IIS if, and only if, one of the associated y-bounds is in the IIS.

Property 2. At most one of each bound is in the IIS – that is, if $x_j \ge L_j$ is in the IIS, then $x_j \le U_j$ is not (same for y-variables).

Suppose Chinneck's IIS code is applied (before entering ANALYZE), say to WOODINFE.MAT, to generate WOODINFE.IIS (these files came with ANALYZE). Then, you can specify `READIN IIS WOODINFE` to read the IIS into blocks. In general, the following row and column blocks are created by this specification.

IIS.LO.......lower bounds in IIS
IIS.UP.......upper bounds in IIS
IIS.LU.......both bounds in IIS.

The last block, IIS.LU, is created when you specify retention of fixed levels, where L=U, such as equations in the case of the row block. Otherwise, if Property 2 is obeyed strictly, only the LO and UP blocks can be created.

Once the blocks have been created with the READIN command, you can interrogate them by copying to the submatrix using the BLOCK command.

6.6 Syntax files.

A syntax file defines the model syntax, which supports the EXPLAIN, SYNTAX and SCHEMA commands and the SYNTAX form of the GRAPH command. Its form is shown in figure 6-5, and you should look at the syntax files that came with ANALYZE, particularly WOODNET.SYN, which we used in chapters 4 and 5.

The syntax file has 3 segments: Sets, Rows and Columns. The Sets segment has no header and must appear first. Each set begins in column 1 and must be an alphanumeric string of 1 to 4 characters. The meaning then follows on the same line, separated by a blank, with up to 64 characters of text.

If set members are declared, they must appear just after the set with a blank in column 1. Each member appears on a separate line with its name, 1 to 4 characters, followed by its meaning (up to 64 characters).

`* This is a comment`	
set 1 [*meaning*]	Segment 1:
member 1 [*meaning*]	Sets
...	
set 2 [*meaning*]	
...	
set s [*meaning*]	
member 1 [*meaning*]	
...	
`ROW SYNTAX`	Segment 2:
row class 1 = *syntax map*	Row syntax
...	
row class r = *syntax map*	
`COLUMN SYNTAX`	Segment 3:
column class 1 = *syntax map*	Column syntax
...	
column class c = *syntax map*	
`ENDATA`	...optional

Figure 6-5. The Form of a Syntax File

The second segment begins with the keyworded line:

ROW SYNTAX

which must be in all CAPS and must begin in column 1. Each row class then follows with a *syntax map*, of the form:

name = *text* [with *set references*].

Usually, the *name* is 1 character, but it can be up to 4 characters. (The objective row is special and need not be defined in this segment.) The *set references* are delimited by the ampersand (&) with the general form:

&name:number[*/number*].

The *name* is the set name, defined in the first segment, and the first *number* (delimited by the colon) is the beginning of the field position in the row's name. If the second number is present, it is the last position in the field; otherwise, the length of the field is the number of characters in the set name.

For example, in WOODNET.SYN you will see the syntax for supply rows defined as shown in the top of figure 6-6. The annotation shows how ANALYZE interprets the syntax map.

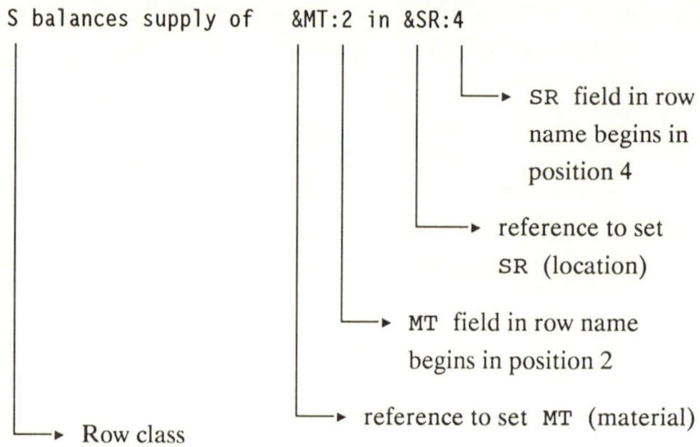

Figure 6-6. A Syntax Map

An equivalent syntax map is:

```
S balances supply of &MT:2/3 in &SR:4/5
```

The set reference, `&MT:2/3`, is the same as `&MT:2` because the length of set MT is two charcters. Similarly, `&SR:4/5` is the same as `&SR:4`. On the other hand, if, instead of naming the set of materials MT, we had named it MATL, the default specification looks for a field of four characters (the length of MATL). If it is only two characters, the set reference must be `&MATL:2/3`.

The third segment begins with the keyworded line:

COLUMN SYNTAX

which, like the ROWS segment, must be in all CAPS and must begin in column 1. Each column's syntax map has the same form as those of the rows.

On-line help for the syntax file is available with the PRINT files SYNTAX.DOC and FILESYN.DOC.

6.7 Rule files.

The keyworded file RULBASE.RUL defines *rule files* that appear as options in the INTERPRT command. You can see the rule files under DOS by entering DIR *.RUL. You should also delete the TEST entry in RULBASE.RUL for production use, as it was included for the ATEST execution. Further, until you are familiar with the use of rule files, you should disable the INTERPRT command by deleting (or renaming) RULBASE.RUL from your production environment. Further, you may modify INTERPRT.DOC for the user to have a dictionary for each option in the INTERPRT command.

A rule file provides a way to prepare responses to the INTERPRT command. You write text in free style, which ANALYZE formats for the response while adhering to the screen settings. To get information about the LP, you either refer to a keyword, or you execute an ANALYZE command. The latter is done on a separate line with a dollar sign ($) in column 1. Keywords can appear anywhere, and they are delimited by the percent sign (%) as a prefix.

There are two types of keywords: character and numeric. Figure 6-7 shows the character keywords. For example, reference to %OBJ substitutes the objective row name. You can reference a substring with (a:b), as in FORTRAN. For example, %OPT(1:3) becomes MIN or MAX.

Key	Meaning
BOUND	bound set name
COLUMN	current column name
OBJ	name of objective row
OPT	sense of optimization (MINIMIZE or MAXIMIZE)
PROBLEM	problem name
RANGE	range set name
RHS	right-hand side set name
ROW	current row name
STATUS	solution status

Figure 6-7. Character Rule Keys

The numeric keywords are either integer or real. The integer keys, shown in figure 6-8, each begin with N. The real keys, shown in figure 6-9, each begin with V.

For example, %NCOLS is substituted by the number of columns in the submatrix, and %VCOLX is substituted by the level of the *current* column. The meaning of *current* will be explained after we describe the commands that you can execute from a rule file.

Key	Meaning
NCBLKS	number of column blocks
NCOLS	number of columns in submatrix
NCSYN	number of column classes in syntax
NESYN	number of entity sets in syntax
NONES	number of ones in submatrix
NONZERO	number of nonzeroes in submatrix
NRBLKS	number of row blocks
NROWS	number of rows in submatrix
NRSYN	number of row classes in syntax
NSTACK	max number in stacks 1 and 2
NSTACK1	number in stack 1
NSTACK2	number in stack 2
NVALUES	number of distinct nonzeroes
NZCOL	number of nonzeroes in column
NZCSUB	number of nonzeroes in column for submatrix rows
NZROW	number of nonzeroes in row
NZRSUB	number of nonzeroes in row for submatrix columns

Figure 6-8. Integer Rule Keys

Key	Meaning
VCOLC	Objective value of current column
VCOLD	Reduced cost of current column
VCOLLO	Lower bound of current column
VCOLUP	Upper bound of current column
VCOLX	Level of current column
VDENSTY	Density (%) in LP matrix
VLOOK	Value from last LOOKUP
VROWC	Objective value of current row (0, except objective)
VROWLO	Lower bound of current row
VROWP	Price of current row
VROWUP	Upper bound of current row
VROWY	Level of current row

Figure 6-9. Real Rule Keys

Figure 6-10 shows the commands recognized from a rule file. The first is the ANALYZE command, which includes session control, like PRINT, and procedural, like BLOCK. The other commands, like LOOP, are particular to rule files. The DEBUG command is for helping to debug the rule file. It pauses at that place and prints whatever comment is specified.

Command	Syntax
ANALYZE	*command spec*
CALC	*param [=] value-expression*
DEBUG	*[comment]*
ENDLOOP	
ENTITY	*member [set]*
EXIT	
GOTO	*label* | TOP
IF	*condition* THEN *command*
INTERPRT	*filespec*
LOOKUP	*key name[, name]*
LOOP	
NEXT	[ROW] [COLUMN]
POP	{1 2} *parameter*
PUSH	{1 2} *string*
QUEUE	{1 2} *string*
SET	*parameter string*
SKIP	*number* (of lines to skip) | LOOP | ENDLOOP
STACK	[{1 2}]
TEXT	[*number* (of lines) [MARGIN=*margin* [*indent*] | CENTER]

Figure 6-10. Rule Commands

All commands must be in all CAPS and appear on a separate line, starting in column 1 as $*command* (see chapter 11).

The EXIT command causes termination of the rule file. If the INTERPRT command is specified with a filespec, another rule file is instantiated, which need not be listed in RULBASE.RUL – that is, it could be *hidden*. Once another rule file is entered, however, its EXIT does not return to this one; it terminates the interpretation. You can, however, use an INCLUDE statement in a rule file (as long as no other INCLUDE file is active).

Do not specify $ANALYZE INTERPRT *filespec*. This will result in an error due to the non-recursive nature of FORTRAN.

To elaborate, the RULBASE.RUL file is interrogated before the interactive session begins (c.f., figure 6-1). The one that came with ANALYZE contains a list of names of rule files. For example, if you look at RULBASE.RUL, you will see PRICE and TEST as 2 of the names. This means files PRICE.RUL and TEST.RUL are rule files, and they appear as user options for the INTERPRT command. We used the PRICE option in chapter 4 for automatic interpretation of prices in an optimal instance of the WOODNET model. The TEST option was used in ATEST when you installed ANALYZE (and should be removed from RULBASE.RUL for production use).

This list of options, which comprises RULBASE.RUL, limits the user to instantiate only those rule files (with the INTERPRT command), but an author of a rule file may instantiate other rule files not listed in RULBASE.RUL. This is done by PRICE.RUL, which first tries some very simple things, such as when the price is zero. Failing this, the PRICE rule file determines if the row is a supply row (S) or a demand row (D) and branches, respectively, to INTERPRT PRICES or INTERPRT PRICED or INTERPRT PRICEB. The rule file, PRICES.RUL, PRICED.RUL, or PRICEB.RUL, respectively, is then instantiated. Control never returns to PRICE.RUL.

The SET rule command allows local variables to be defined and [re-]set. When referenced in the text or another command, these are delimited by 2 percent signs (%%). For example, you can SET NR = %NROWS. This will create your variable, called NR, which is set equal to the system key value NROWS – that is, the number of rows in the submatrix. Later, even if the submatrix changes, the value of %%NR is what you set it to be originally. You can also use the same keyword: SET NROWS = %NROWS. Your variable, now called NROWS, is distinct from the system keyword because all references to it have the form %%NROWS instead of %NROWS.

All your variables are treated as strings, but some numeric calculation and comparison is possible. The CALC command allows a simple addition, like CALC NROWS = %%NROWS + 1. String comparisons are simply equal (=) or not equal (< >), but you can specify the numeric comparisons (<, < =, >, > =) as well. If the 2 strings being compared are not numeric, these relations are undefined, which causes an error message to appear. If, however, the string represents a numeric value, the comparison will not yield an error. Care must be taken with string comparisons; for example, the strings 1.0 and 1 are not the same.

Branching can be unconditional with the GOTO and SKIP commands. The former branches to a label in the rule file; the latter skips a number of lines. A label appears, as in a DOS BAT file, on a separate line with a colon (:) in column 1.

Unlike DOS, however, branching with the GOTO command must be in a forward direction, except within a loop.

The LOOP command loops over the submatrix. It ends with a NEXT command, which can specify ROW or COLUMN or both. The last is useful to loop down the diagonal of a square submatrix, like the basis. This is what the reference to the *current row* and *current column* mean. The keywords %ROW and %COLUMN are the names of the row and column, respectively, determined by the pointer to the submatrix. Initially, these are the first ones. Only with the NEXT command, which must follow a LOOP command, does this pointer change. Nested loops are not permitted. The ENDLOOP statement branches to the line immediately following the NEXT statement.

The GOTO command has the special branching specification GOTO TOP, which means it goes to the top of the rule file. The SKIP command permits SKIP LOOP to skip to the NEXT statement of the loop. If the NEXT ROW statement is reached, and the row pointer is at the last row in the submatrix, the loop ends (similarly, for NEXT COLUMN). It also permits SKIP ENDLOOP to branch to the statement following the end of the loop (thus terminating the loop).

Each command must be fully specified (not abbreviated), but the options following the command can be abbreviated. In particular, the ANALYZE command can be followed by any valid abbreviation of an ANALYZE command specification. Keywords and parameters cannot be abbreviated in their references.

String substitution is performed as the rule file is read and interpreted, and the settings from the main command level are still in effect. This could also come from an execution file, which then instantiates the rule file. If the rule text requires the string delimiter to be a literal, it must be doubled. For example, using the single quote (') as a string delimiter (the default), 's is parsed as a string whose name is s, whereas ''s is the literal 's.

On-line help for the rule file is available with the PRINT file RULEBASE.DOC, which includes the tables of keywords and commands. You might also see the rule files that came with ANALYZE (they have the suffix .RUL). Further descriptions and examples are in chapter 11 of this User's Guide.

6.8 Submatrix, block, schema and graph files.

The SUBMAT SAVE (direct or in the sub-command level) allows you to save your submatrix. This is done by creating an execution file that enters the rows and columns, in the submatrix saved, with the SUBMAT command. You can edit this SUBMAT file, and it can be used in different instances of the model.

Because the execution file generated uses the SUBMAT prefix and suffix, you may need to fully qualify it in the EXECUTE command. For example, suppose a submatrix is saved into WOODNET.SUB. Then, you must EXECUTE WOODNET.SUB (not WOODNET.EXC). The execution file contains a _PAUSE to remind the user that it will merge with the current submatrix. To have precisely the saved submatrix reconstructed, the user must first clear the submatrix.

Blocks can be saved by specifying BLOCK SAVE *filespec*. This is a binary file whose _SETUP type is BLOCK. All blocks are saved; and, BLOCK LOAD *filespec* can be used to load that set of blocks (which replaces the current blocks).

The schema can also be saved with SCHEMA SAVE *filespec*. This is usually not necessary because the construction for the schema display is fast, but it may be desirable to do so for exportation to a slow computer. The SCHEMA LOAD specification enables the SCHEMA DISPLAY, even without reading a syntax file into memory.

Another view is the graph of the submatrix, which is obtained by the GRAPH command. The SAVE and LOAD specifications enable saving graphs of different submatrices (perhaps from within a rule file), and loading them later. This is a binary file (unformatted), whose _SETUP type is GRAPH.

6.9 Setting up an environment for analysis.

Now that we have described each of the file types that can be used for enhanced environment control and for linking to other systems, consider combinations to setup your own environment for analysis. The setup may be for industrial use, bound to instances of a particular model, or it may be for educational reasons. The latter case is what we did in preparing exercises for this User's Guide, and we want to extend this to a more general laboratory environment. With the MODLER and GAMS links, this is particularly well suited to complement a linear programming course with students required to solve some analysis problems, such as those we did in chapters 4 and 5 (with more to come in chapter 7).

Recall the special files in figure 6-1. In particular, begin with a PROFILE.EXC that introduces the student (or practicing analyst) to ANALYZE. Here is where you can not only set environment controls, but also you can print a short message, such as is done in WOODNET.EXC (see, also, the other EXC files for tips). The PROFILE can give a dictionary of exercises. An example is in EXERCISE.PRO, where we have gone through some of them in this User's Guide (you might try the others, like PRSUB).

Prepare LP instances by creating matrix and solution files (or with the LP file described in §8.2, which is a link with GAMS). Define your names so they can be translated with a syntax file, which you also create (GAMS does this automatically, and MODLER has a command to write a syntax file). Further prepare the packed file for the instance by using the BASIS command to setup the basis in order to support the RATEOF command. Finally, use the WRITEOUT command to save the packed file.

After creating a library of packed files and syntax files, which are bound to models, you can create execution files to read them and print a brief message to get the student/analyst started. You might also add some documentation (PRINT) files and dictionaries to offer more guidance.

Begin with this paradigm to develop simple exercises, such as the ones we did in chapters 4 and 5. In some instances, you might prepare rule files. For educational purposes, however, you can ask the students to create their own rule files. This stretches their knowledge of both linear programming analysis and of presenting results to non-experts. Once their rule files run well on the prepared problems, you can run them for new instances to see how they do. By having to go through the steps of analysis, the students are forced to think hard about their rules and about presentation of results. This can be very effective for them to learn with problems of modest (but nontrivial) sizes, then tackle more complex cases encountered in practice.

For anomalies, such as infeasible instances, the students can learn more about the use of Phase 1 prices and their limited information value. For non-network models, the students can learn more about underlying economic structures, such as competition for resources that might be reflected in the spike structure of the basis.

The REDUCE command may give reductions that depend on the instance or that result from model structure, apart from data. One exercise is to ask the student to determine the reasons for the reductions that were found, particularly fixed activity levels. This involves a traceback using the reductions report and interactive navigation in ANALYZE.

It is likely that the world of ANALYZE will expand with future versions incorporating innovative use of this type. Please let the author know of such experiences, good or bad (see cover for address).

Chapter 7. Advanced Exercises

Modeling is about insight, not numbers.
Arthur M. Geoffrion

In this chapter we illuminate some of the added capabilities and limitations of the ANALYZE procedures. First, we explore a forestry model and reinforce what we have done in previous chapters. Then, we consider an infeasible instance of the same model for which the procedures we used in chapters 4 and 5 give little help to diagnose the cause of the infeasibility. Second, we describe an anomaly with the REDUCE algorithm, which could give a false detection of infeasibility in practice. Third, we describe some additional ways ANALYZE could help with model management.

7.1 A forestry model.

We explore a forestry model, beginning with an optimal instance. Figure 7-1 shows the result of executing the FOREST.EXC execution file. This read both the packed file (matrix + solution) and the syntax file into memory. Then, as shown in figure 7-2, we specify the SUMMARY command to see its LP statistics.

```
ANALYZE ... EXEC FOREST
  FOREST.....an optimal forestry model.
     You now have an optimal model + the syntax file.  Analyze
     the properties, beginning with model attributes.  Use
     query commands and SYNTAX to help you.  Then, try using
     the TRACE and RATEOF commands.  Try to explain what you
     learn about the model, using the EXPLAIN and SYNTAX
     commands to put your explanation in English.  After this,
     there is an infeasible version of the same model, and you
     are to diagnose the problem.  When ready, READ P FOREST6
     This will load the infeasible version into memory.  To
     obtain the syntax (for EXPLAIN), READ S
```

Figure 7-1. Loading a Forestry Model into Memory

```
ANALYZE ... SUM
Problem FOREST is to MINIMIZE COST
RHS = RHS  RANGE = none  BOUND = LAND
Source file: FOREST5.PCK
Syntax file: FOREST.SYN
Solution source: MINOS
Solution Status: OPTIMAL
LP Statistics:
  56 rows (1 free) and 80 columns (0 fixed)
  225 nonzeroes (48 distinct... 125 ones)
  Density= 5.02232%  Entropy= 21.3333%  Unity= 55.5556%
Row    D1        has the most nonzeroes = 5.
Column HBB1      has the most nonzeroes = 3.
```

Figure 7-2. Summary Statistics of a Forest Model Instance

The summary statistics and some additional information can be seen with the rule file STAT.RUL by specifying INTERPRT STAT, as shown in figure 7-3. The advantage of this display is that it avoids some of the LP jargon, given by the SUMMARY command, and it gives some additional information, namely statistics for the embedded netform, degeneracy, and the basis. (The term 'vanilla', which is what is in the rule file, is a bit of LP jargon that means $x \geq 0$.)

Discrepancy Note

> Here and elsewhere, the screen width was first reduced to 70 in order to fit in this User's Guide (then restored to 79).

The embedded netform was obtained by the BLOCK command. The degeneracies were obtained by simple conditional delineation, where a primal variable is degenerate if it is basic, but its level equals one of its bounds. A dual degeneracy corresponds to a non-basic primal variable with zero price. You might examine the file STATS.RUL to tailor the output for your environment (see, also, MODEL.RUL).

```
ANALYZE ... INTERPRT STAT
                      Statistical Summary of FOREST

                               Dimensions
                               ~~~~~~~~~~
      56 Rows.......... 25 Equations, 30 Inequalities, 1 Free
      80 Columns....... 75 Vanilla, 5 Bounded, 0 Fixed
      225 Nonzeroes..... 48 Distinct, 125 ones, Density = 5.02232

      Embedded Netform is 56 x 60:
   ...0 Production, 5 Consumption, 55 Transfer activities.

                           Solution Statistics
                           ~~~~~~~~~~~~~~~~~~~
  Objective: MIN COST = 285837
  Primal Degeneracies: Rows = 3 Columns = 21
  Dual Degeneracies: Rows = 0 Columns = 0
  The basis has 52 binding rows (and 52 basic structurals) with 118
  nonzeroes.
```

Figure 7-3. Using the INTERPRT Command to See Statistics for the Forest Model Instance

We next get syntax information to gain some initial understanding of the Forest model's structure. In figure 7-4 we specify the SCHEMA and SYNTAX commands, respectively, for this purpose.

```
ANALYZE ... SCH D TABLE, DOMAIN
        :::::::::::::::::: BLOCK SCHEMA ::::::::::::
         I(AR,T)      H(AR,T)     W(AR,T)     RHS
 B(AR,T) -1/1         -1          -1          = 0
 D(T)                 0.2/0.4                 >=1050/1250
 W(AR,T) -0.75/-0.25               1          >=0
 COST                 5.90/13.5   1.18/4.5    ...MIN
 LAND:LO 0            0           0
 LAND:UP 10000/*      *           *

    DOMAIN INFORMATION
  AR      area
  T       time period
ANALYZE ... SYN R,C

Row syntax has 3 classes
  A row that begins with B balances land use in some area in some
      time period.
  A row that begins with D demands timber in some time period.
  A row that begins with W requires percent of wilderness in some
      area in some time period.

Column syntax has 3 classes
  A column that begins with I carries inventory of land in some area
      at end of some time period.
  A column that begins with H harvests timber in some area in some
      time period.
  A column that begins with W preserves wilderness in some area in
      some time period.
```

Figure 7-4. Getting Syntax Information About the Forest Model

Eventually, we shall want to know the sets that comprise the domains shown in the schema in figure 7-4. One way to do this is with the SYNTAX command, specifying the SETS (or simply S) option, as shown in figure 7-5. (You can specify SYN S AR to see just the set of areas. Try it.)

```
ANALYZE ... SYN S
  2 sets...
   Set AR = area
            BR = Blue Ridge
            SM = South Meadows
            BB = Black Butte
            NR = North Ridge
            GA = Glenn Acres
   Set T = time period
            0 = initial period
            1 = period 1
            2 = period 2
            3 = period 3
            4 = period 4
            5 = period 5
            6 = period 6
            7 = period 7
            8 = period 8
            9 = period 9
```

Figure 7-5. Using the SYNTAX Command to See the Sets

To increase our familiarity with the Forest model, and to reinforce what was presented in previous chapters, consider the demand prices, displayed in figure 7-6. We see this instance has 5 periods with demand rows D1,...,D5. The demand requirements are modeled as lower bounds on the demand rows (rather than with activities, as in WOODNET).

```
ANALYZE ... D R D
5 Rows are in submatrix.
ROW      STAT  LEVEL (Y)     LO_BOUND       UP_BOUND      PRICE (P)
-----------------------------------------------------------------
D1        L      1050.000     1050.000          *             46.903
D2        L      1100.000     1100.000          *             57.900
D3        L      1150.000     1150.000          *             76.110
D4        L      1200.000     1200.000          *            101.907
D5        L      1250.000     1250.000          *            139.443
```

Figure 7-6. Displaying the Demand Rows

Figure 7-7 lists the equations after putting all columns into the submatrix. We see that demand requirements are fulfilled by the Harvest activities from different areas (c.f., figure 7-4). For practice, we shall interpret the second period's demand price of $57.90 (shown in figure 7-6).

```
ANALYZE ... SUB CO *
ANALYZE ... L
   1050    <= D1          =   0.3 HBB1 +  0.4 HBR1 +  0.3 HGA1 +  0.4 HNR1
                          +   0.2 HSM1
   1100    <= D2          =   0.3 HBB2 +  0.4 HBR2 +  0.3 HGA2 +  0.4 HNR2
                          +   0.2 HSM2
   1150    <= D3          =   0.3 HBB3 +  0.4 HBR3 +  0.3 HGA3 +  0.4 HNR3
                          +   0.2 HSM3
   1200    <= D4          =   0.3 HBB4 +  0.4 HBR4 +  0.3 HGA4 +  0.4 HNR4
                          +   0.2 HSM4
   1250    <= D5          =   0.3 HBB5 +  0.4 HBR5 +  0.3 HGA5 +  0.4 HNR5
                          +   0.2 HSM5
```

Figure 7-7. Listing the Demand Equations

As before, we begin by tracing row D2 and basic columns, as shown in figure 7-8. (Because we know there is no dual degeneracy (c.f., figure 7-3), we need not distinguish the status specification, S=B, from the reduced cost specification, D=0.) Then, we picture the result.

```
ANALYZE ... TRACE D2,* S=B
 TRACE complete...submatrix is set
ANALYZE ... PIC
                              Picture of submatrix
         H I I W W
         G G G G G
         A A A A A
         2 0 1 1 2
BGA1       + - -   =  0
BGA2     -   +   - =  0
COST     +     + + = MIN
D2       +         >  +
WGA1       -   +   >  0
WGA2         -   + >  0
```

Figure 7-8. Beginning the Analysis to Understand the Demand Price in Period 2 (Row D2)

To help understand this submatrix, we use the EXPLAIN command to see a syntactic translation of each row and column. This is shown in figure 7-9.

```
ANALYZE ... EXP R
  Row BGA1 balances land use in Glenn Acres in period 1.
  Row BGA2 balances land use in Glenn Acres in period 2.
  Row COST is the objective (MINIMIZE).
  Row D2 demands timber in period 2.
  Row WGA1 requires percent of wilderness in Glenn Acres in period 1.
  Row WGA2 requires percent of wilderness in Glenn Acres in period 2.
ANALYZE ... EXP C
  Column HGA2 harvests timber in Glenn Acres in period 2.
  Column IGA0 carries inventory of land in Glenn Acres at end of
    initial period.
  Column IGA1 carries inventory of land in Glenn Acres at end of
    period 1.
  Column WGA1 preserves wilderness in Glenn Acres in period 1.
  Column WGA2 preserves wilderness in Glenn Acres in period 2.
```

Figure 7-9. Explanations of Rows and Columns in the Submatrix

From figure 7-8, we do not see a simple path from supply to demand, so the analysis will be more complex than that of network models (like the exercises in the previous chapters for the WOODNET model). The columns in the submatrix are basic (use DISPLAY to see this for yourself), and we proceed by listing the equations, shown in figure 7-10.

```
ANALYZE ... L
      0     = BGA1    =  IGA0 - IGA1 - WGA1
      0     = BGA2    =- HGA2 + IGA1 - WGA2
      MIN     COST    =   12.15 HGA2 +  1.8 WGA1 +  1.62 WGA2
   1100    <= D2      =   0.3 HGA2
      0    <= WGA1    =  WGA1 -  0.5 IGA0
      0    <= WGA2    =  WGA2 -  0.5 IGA1
```

Figure 7-10. Listing the Submatrix Equations

Recalling that the activities in the submatrix are basic, we can write the dual (price) equations:

```
From HGA2:   .3P(D2)    - P(BGA2)                          = C(HGA2)
From IGA0:    P(BGA1)            - .5P(WGA1)               = 0
From IGA1:   -P(BGA1) + P(BGA2)             -.5P(WGA2)     = 0
From WGA1:   -P(BGA1)            +   P(WGA1)               = C(WGA1)
From WGA2:              - P(BGA2) +               P(WGA2)  = C(WGA2)
```

where P(*row*) is the price of *row* and C(*column*) is the objective coefficient (COST) of *column*. Try LIST //DUAL to get the same information (in a slightly different form).

We begin the price interpretation with the following.

> The price of row D2 ($57.90) is the cost of harvesting ($12.15 from C(HGA2)) plus the price of land in Glenn Acres in period 2 - that is, <*explain* P(BGA2)>. For each unit of timber harvested, however, only .3 reaches the demand, so the total price = (12.15 + 5.22)/.3 = 57.90.

This used the first price equation (from activity HGA2). To complete the interpretation, we must explain the price of the balance row, P(BGA2), which is indicated in this template by <*explain* P(BGA2)>. We could simply put in its value ($5.22), but this naturally raises the question as to how it is determined.

This can be interpreted with the remaining price equations, but the case is more complex than WOODNET, which had a hierarchical ordering due to its being a network model. Now there is no single equation to explain P(BGA2) in terms of the original coefficients. Instead, we must consider price equations simultaneously.

We can, however, decompose the system to require explanations of only 2×2 subsystems. To do this, we first organize equations, as shown in figure 7-11. The rows of this matrix are the basic activities, which comprise the source of the price equations. The columns of this matrix are the row prices we want to interpret. The location of D2 at the rear, from HGA2, reflects what we already said: The price of row D2 is determined once we determine P(BGA2). (In a network model, such a matrix would be triangular, so there would be a simple path that comprises a causal chain in the reasoning that results in the interpretation.) All 5 row prices are involved in the final interpretation, but we can proceed with the rearranged matrix from the Northwest to the Southeast.

	WGA1	BGA1	WGA2	BGA2	D2
IGA0	-.5	1			
WGA1	1	-1			
IGA1		-1	-.5	1	
WGA2			1	-1	
HGA2				-1	.3

Figure 7-11. A Rearrangement of the Price Equations

The first 2×2 subsystem solves for P(WGA1) and P(BGA1):

From IGA0: P(BGA1) - .5P(WGA1) = 0
From WGA1: -P(BGA1) + P(WGA1) = C(WGA1) = 1.80

This has the solution, P(BGA1)=1.80 and P(WGA1)=3.60. Then, the second 2×2 subsystem solves for P(WGA2) and P(BGA2):

From IGA1: P(BGA2) -.5P(WGA2) = P(BGA1) = 1.80
From WGA2: - P(BGA2) + P(WGA2) = C(WGA2) = 1.62

This has the solution, P(BGA2)=5.22 and P(WGA2)=6.84.

The problem now is to put these facts together in giving an interpretation of the price of the demand row (D2). The interpretation has a *causal decomposition* in that we can explain P(D2) once we interpret P(BGA2) – that is, we simply then add C(HGA2) and divide by .3, which can all be interpreted in natural language form.

We can interpret P(BGA2) as part of a 2×2 subsystem once we can interpret P(BGA1). Again, we backup and find that we can complete the interpretation with another 2×2 subsystem. How to explain a 2×2 subsystem is not entirely clear; it generally depends on the model. This can become more complicated in other models, where one must interpret a larger subsystem – that is, where there is no further causal decomposition in an m×m subsystem for large m. All prices in an *irreducible* subsystem are inter-dependent. Algebraically, this is no problem; we know how to solve large systems of linear simultaneous equations. There is, however, some art in forming a final interpretation.

In this instance, we can form the following interpretation.

```
The price of row D2 ($57.90) is the cost of harvesting ($12.15)
plus the price of land in Glenn Acres in period 2 ($5.22).  For
each unit of timber harvested, however, only .3 reaches the
```

```
demand, so the total price = (12.15 + 5.22)/.3 = 57.90.  The
harvesting cost is specified (input), so it remains to interpret
the price of land in Glenn Acres in period 2.

The price of land in Glenn Acres in period 2 depends on the cost
of wilderness allocation, which is specified (input) = $1.62,
and on the price of land in Glenn Acres in period 1, which is
determined to be $1.80 (see next paragraph).  The price of the
wilderness constraint is inseparable from the price of land in
Glenn Acres in period 2 due to the associated activities that
allocate wilderness and carry inventory of land in Glenn Acres
from period 1 to period 2.  (The exact form of this
interdependence can be seen by picturing the columns of
activities IGA1 and IGA2.)

Finally, it remains to interpret the price of land in Glenn
Acres in period 1, which is the price of row BGA1 = $1.80.  This
price depends on the cost of wilderness allocation, which is
specified (input) = $1.80.  The activity that carries Glenn
Acres land (not allocated for wilderness or for harvesting)
implies a price relation:  the price of Glenn Acres land in
period 1 = .5 times the price of the Glenn Acres wilderness
constraint in period 1.  In addition, the two prices must differ
by the cost of wilderness allocation ($1.80).  These two price
relations, implied by the associated Inventory and Wilderness
allocation activities (IGA0 and WGA1, respectively), yield the
resulting price.
```

Paragraphs 2 and 3 are worded differently to illustrate some variation in interpreting a 2×2 subsystem, as part of a final interpretation that is as natural as possible. It is this kind of response formation that can be put into a rule file to support analysis by users who might not understand linear programming algebraic structures.

Now consider another use of blocking. Figure 7-12 begins by erasing blocks, which were created by the STAT rule file (cf., figure 7-3) to obtain embedded network statistics. Then, as shown in figure 7-12, the COUNT command sets the submatrix to have demand rows (i.e., whose name begins with D) and harvest activities (i.e., those whose name begins with H) with positive activity levels. Knowing that the activities have the set Area in their domains, whose members are in character positions 2&3 in their names, we block by syntax to partition the columns by area. The demand rows are partitioned by time period, which has the set T in their character position 2. When we specify the block picture, we see which area fulfills demand in each period.

```
ANALYZE ... BL R E *
ANALYZE ... BL C E *
ANALYZE ... C D,H X=1/*
  Submatrix has  5 rows, 7 columns, and 7 nonzeroes
ANALYZE ... BL C S *??
           3 block(s) added
ANALYZE ... BL R S *?
           5 block(s) added
ANALYZE ... BL P
              BLOCK PICTURE
       S S S
       Y Y Y
       N N N
       . . .
       B G N
       B A R
SYN.1      +
SYN.2    +
SYN.3    + +
SYN.4      +
SYN.5  +   +
```

Figure 7-12. Blocking by Syntax to see Distribution Pattern

The block names begin with SYN. followed by the class corresponding to the query characters. The column classes, for example, are *BB, *GA and *NR. You can use the SYNTAX SETS command to see what these mean: BB represents Black Butte, GA represents Glen Acres and NR represents North Ridge. These are forests in the U.S. We see, in the block picture, Black Butte harvests to fulfill demand in period 5; Glenn Acres (GA) harvests to fulfill demand in periods 2 and 3; and, North Ridge harvests to fulfill demand in all periods, except period 2.

Now we turn our attention to an infeasible instance of the Forest model. Figure 7-12 shows the first step of reading the packed LP file (matrix + solution), then the syntax file, into memory. You might use the SUMMARY, INTERPRT STATS, SCHEMA, and SYNTAX commands (c.f., figures 7-2 to 7-4) to see that this instance has the same structure. You might also try the REDUCE command, and you will see that the infeasibility is not detected. Our problem is to diagnose a *probable cause.*

```
ANALYZE ...REA P FOREST6
 Packed LP read from FOREST6.PCK

ANALYZE ...REA S
 Syntax read from FOREST.SYN
ANALYZE ...D R * S=I
 4 Rows are in submatrix.
 ROW     STAT  LEVEL (Y)    LO_BOUND      UP_BOUND      PRICE (P)
 --------------------------------------------------------------------
 WBR2     I  -  2253.704        0             *             1.0
 WBR3     I  -  2253.704        0             *             1.0
 WBR4     I  -  2253.704        0             *             1.0
 WBR5     I  -  2253.704        0             *             1.0
ANALYZE ...D C * S=I
 0 Columns are in submatrix.
```

Figure 7-12. Reading an Infeasible Instance of the Forest Model and Displaying the Infeasibilities

As shown in figure 7-12, there are 4 infeasible rows and no infeasible columns. We begin by exploring the first infeasible row, namely WBR2.

In figure 7-13 we picture the row and find its negative level could be increased by the associated wilderness allocation activity, namely WBR2. (Although they have the same name, *row* WBR2 is distinct from *column* WBR2.) It must be that increasing this activity worsens the total sum of infeasibility because that is the meaning of a Phase I solution. To see the secondary effects of increasing the associated wilderness activity, we enter all rows into the submatrix, and picture the result. The negative coefficient in row BBR2 of column WBR2 must be why its level cannot be increased to reduce the infeasibility of row WBR2, but we must investigate further.

```
ANALYZE ... PI WBR2,*
                                  Picture of submatrix
          I W
          B B
          R R
          1 2
WBR2      - + >  0
ANALYZE ... PI *
                                  Picture of submatrix
          I W
          B B
          R R
          1 2
BBR1      -   =  0
BBR2      + - =  0
COST        + = MIN
WBR2      - + >  0
```

Figure 7-13. Beginning an Investigation Into Infeasible Row WBR2

In figure 7-14 we first delete the objective row (COST) from the submatrix, then add all basic columns. We see that an increase in the level of the wilderness allocation activity (WBR2) directly affects the row that balances land in Blue Ridge in period 2 (namely, BBR2). From the sign pattern in the picture, we see that to balance the equation, we must either increase the level of inventory in the previous period (namely, activity IBR1) or decrease the level of inventory in this period (namely, activity IBR2). Either change has tertiary effects, which require further analysis.

```
ANALYZE ... SUB D R C
ANALYZE ... SUB A C * S=B
ANALYZE ... PI
                            Picture of submatrix
        H I I W W
        B B B B B
        R R R R R
        1 1 2 1 2
BBR1    - -   -   =  0
BBR2      + -   - =  0
WBR2      -     + >  0
ANALYZE ... L
     0     = BBR1     =- HBR1 - IBR1 - WBR1
     0     = BBR2     =  IBR1 - IBR2 - WBR2
     0    <= WBR2     =  WBR2 -  0.75 IBR1
```

Figure 7-14. Seeing the Direct and Secondary Effects of Increasing the Wilderness Allocation Activity (WBR2)

Since the columns in the submatrix at this point are basic, except the original wilderness allocation activity (WBR2), their levels can change to maintain the equations. In particular, let θ denote a marginal increase in the level of the wilderness activity – that is, $\theta = \delta X$(WBR2). The equations in figure 7-14 require that the compensating changes satisfy:

From BBR1: $\delta X(HBR1) = - \delta X(IBR1) - \delta X(WBR1)$

From BBR2: $\delta X(IBR2) = -\theta + \delta X(IBR1)$

From WBR2: $\delta Y(WBR2) = \theta - .75\delta X(IBR1)$

As we continue this line of exploration, the system can become untenable. The indirect effects are already showing up because the change in inventory, $\delta X(IBR1)$ (if positive), reduces the increase in row WBR2, and it affects other levels in the other two equations. If we proceed this way, we would see effects on the other infeasible rows. Fortunately, there is a shortcut to see all of the effects of changing the level of the wilderness activity (WBR2).

Consider using the RATEOF command, as in figure 7-15, to see why the level of activity WBR2 cannot be increased without violating other infeasibilities worse than it improves the one we investigated – that is, we want to see why the net effect of increasing the level of activity WBR2 is to increase the total sum of infeasibility despite its positive effect on the infeasible row (WBR2).

```
ANALYZE ... RAT C WBR2 //E
  Rates of substitution for COL WBR2
  ...Status=L Level=0  Price=0.0376700
 IBR1      =     3004.939  +         1.095 x WBR2
 IBR2      =     3004.939  +   0.0958904   x WBR2
 IBR3      =     3004.939  +   0.0958904   x WBR2
 IBR4      =     3004.939  +   0.0958904   x WBR2
 HBR5      =     3004.939  +   0.0958904   x WBR2
 IGA1      =    24160.082  -         1.461 x WBR2
 IGA2      =     8413.374  -   0.7305935   x WBR2
 IGA3      =      373.353  -   0.3652968   x WBR2
 HGA4      =      186.676  -   0.1826484   x WBR2
 INR4      =     4080.007  -   0.1369863   x WBR2
 COST      =   457198.718  +        10.961 x WBR2
 HGA1      =      839.918  +         1.461 x WBR2
 HNR4      =     1922.492  +   0.1369863   x WBR2
 HNR5      =      120.060  -   0.0958904   x WBR2
 WBR2      = -   2253.704  +   0.1780822   x WBR2
 WBR3      = -   2253.704  -   0.0719178   x WBR2
 WBR4      = -   2253.704  -   0.0719178   x WBR2
 WBR5      = -   2253.704  -   0.0719178   x WBR2
 WGA2      =    12080.041  -   0.7305935   x WBR2
 WGA3      =     4206.687  -   0.3652968   x WBR2
 WGA4      =      186.676  -   0.1826484   x WBR2
 WNR5      =     1224.002  -   0.0410959   x WBR2
 HBR1      =     1995.060  -         1.096 x WBR2
 ------------------------------------------------
       WBR2      Range:       - 574.820     1022.056
       Blockage:              HGA1          IGA3
```

Figure 7-15. The Rates of Substitution for the Wilderness Allocation Activity WBR2

Figure 7-15 shows the rates of substitution with the levels of the basic variables. In particular, it affects all of the infeasible rows! As before, let $\Theta=\delta$X(WBR2). Then, we see:

```
δY(WBR2)  =  +   0.1780822 θ
δY(WBR3)  =  -   0.0719178 θ
δY(WBR4)  =  -   0.0719178 θ
δY(WBR5)  =  -   0.0719178 θ
              ==============
 Net Rate =  -   0.0376700     ——▸ - Reduced Cost (D)
             |
             └——▸ Infeasible row levels become more infeasible
```

Seeing the rates of substitution is a shortcut to seeing the effects of changing a nonbasic variable's level in order to reduce the infeasibility of a row (WBR2 in this case). As we began to see previously, the immediate positive effect on the level of row WBR2 is reduced due to the necessary change in the inventory level. The RATEOF command lets us see how much affect this has on which (basic) levels.

To make the explanation easier, we could change our approach and consider variables at upper bound. The column display, shown in figure 7-16, reveals that all of the initial inventory levels are at their upper bound. From this we may suspect the problem is that there is not enough land to meet the demand requirements. In particular, IBRO is at its upper bound (20000). The negative sign of its reduced cost suggests we can reduce the infeasibility by increasing this upper bound – that is, make more land available initially in Blue Ridge.

```
ANALYZE ... D C * S=U
5 Columns are in submatrix.
COL      STAT  LEVEL (X)     LO_BOUND      UP_BOUND      PRICE (D)
-----------------------------------------------------------------
IBBO      U     10000.000        0          10000.000   - 0.6269300
IBRO      U     20000.000        0          20000.000   - 0.0719200
IGAO      U     50000.000        0          50000.000   - 0.1078800
INRO      U     25000.000        0          25000.000   - 0.5525600
ISMO      U     30000.000        0          30000.000   - 0.0719200
```

Figure 7-16. Displaying the Submatrix Columns to See Their Levels and Bounds

Based on this information, we might be tempted to declare the cause to be insufficient land. On the surface this is plausible because its (Phase I) reduced cost indicates that it does affect the sum of infeasibility. The real cause, however, is deeper.

In either case, whether we diagnose the infeasibility to be insufficient land or difficulty with wilderness allocation, the net result is still only a *possible* explanation. In this example, the infeasibility is more difficult to diagnose. The real cause of infeasibility is not possible to surface without seeing the model description or enough instances to reveal the relevant information. The demands are determined by an initial set of values plus a growth rate. The wilderness constraints represent percentages of land inventory that must be allocated. These two properties of the formulation conspire to ensure infeasibility for some horizon. That is, in this model, due to the way it was formulated, there will always be a threshold horizon, below which the instance is feasible and above which it is infeasible. Realizing that is beyond the purpose of this User's Guide, we leave it at that. The point is that some infeasibilities can defy correct diagnosis by analyzing only one instance of the model. The identification of initial inventory as a prime candidate for the diagnosis is reasonable, but it is incorrect. Our purpose in having gone through this exercise was both to reinforce the understanding of how to use ANALYZE and to illustrate its limitations.

Now consider the use of the AGGREGAT command, which computes the linear combination of rows in the submatrix weighted by the resident dual prices. As illustrated in chapter 5, the phase I prices of an infeasible LP can be used to construct an aggregate infeasibility. Let us apply this to FOREST6.

Figure 7-17 shows the aggregate constraint for this infeasible instance. Unlike the simpler case of WOODINFE (c.f., figure 5-48), there are a great many coefficients in this aggregate constraint. The display offers some help, but tracing is still needed for a comprehensible diagnosis. All we know is that this implied equation (with the aggregate coefficients) is inconsistent with the bounds on the activities. The aggregate range requires a solution whose (aggregate) level is at least 38087.816, but the bounds imply that this level cannot exceed 29073.002.

```
ANALYZE ...   SUB *
ANALYZE ...   AGG R
  Aggregate ROW range:    38087.816 to         *
  Coefficient:                          Range
  - 0.6201560     HBB1               -*                0
  - 0.6830320     HBB2               -*                0
  - 0.6230470     HBB3               -*                0
     (many more coefficients are displayed)
     -   2.818    WBR6               -*                0
          ...Myopic range:        - *      to     29073.002
```

Figure 7-17. Applying the AGGREGAT Command to FOREST6

Anomalies arise when the optimizer does not produce a pure phase I solution. This can happen because the simplex-based linear programming optimizers do not solve a pure phase I problem. If only row variables are infeasible, however, and if no special scaling was used, you should have a phase I solution. If column levels are infeasible, these are called *migrating infeasibilities*, and anomalies could appear with the aggregation.

This did not happen, and the aggregate constraint is infeasible. Other anomalies could arise from arithmetic errors. For example, the coefficient of an unbounded activity could compute to a small value, even though its theoretical value is zero (e.g., all basic variables have 0 coefficient if true phase I prices are resident). In this case, the myopic range could appear as infinite due to failure to recognize the zero. A tolerance is used to prevent this, but that too can cause anomalies when a nonzero coefficient is considered to be zero due to being within the tolerance.

7.2 A REDUCE anomaly.

One of the small test problems is the matrix file REDTEST.MAT. The first 2×2 portion poses an anomaly for the REDUCE algorithm, and it may cause difficulties in real cases. The LP equations are:

$$\tfrac{1}{2} x_1 + x_2 = 1$$
$$x_1 + x_2 = 2$$

where $(x_1, x_2) \geq 0$. This system has the unique solution $x = (2, 0)$, and it is this uniqueness that causes a problem.

Immediately, the first equation sets finite upper bounds of 2 and 1, respectively. With $x_2 \leq 1$, the second equation strengthens the first lower bound to 1. At a general iteration, say k, the variables have intervals of the form $x_1 \in [L^k, 2]$, where $L^k < 2$, and $x_2 \in [0, U^k]$, where $U^k > 0$. In fact, a closed form expression for the successive bound reduction is given by:

$$L^k = 2 - (\tfrac{1}{2})^{k-1} \text{ and } U^k = (\tfrac{1}{2})^k \text{ for } k > 1$$

for the k-th iteration (see §9.6).

The convergence to the unique solution with successive bound reduction is therefore not necessarily a finite algorithm. There is a tolerance, say τ, such that a variable is fixed if its upper and lower bounds are within τ of each other. Applying this to x_1 and x_2, we have:

Fix $x_1 \in [L^k, 2]$ if $(\tfrac{1}{2})^{k-1} \leq \tau$; equivalently, if $k-1 \geq -\log_2 \tau$.

Fix $x_2 \in [0, U^k]$ if $(\tfrac{1}{2})^k \leq \tau$; equivalently, if $k \geq -\log_2 \tau$.

If we iterate to k = $\lceil -\log_2 \tau \rceil$, we fix x_2 (before x_1). Suppose we fix x_2 at its mid-point, $(\frac{1}{2})^{k-1}$. Then, the first equation holds, but equation 2 now has the range,

$$[2 - (\tfrac{1}{2})^k + (\tfrac{1}{2})^{k-1},\ 2 + (\tfrac{1}{2})^{k-1}].$$

The least value is therefore $2 - (\frac{1}{2})^k + (\frac{1}{2})^{k-1} = 2 + (\frac{1}{2})^k$. The error is $(\frac{1}{2})^k$, but it may be enough to cause a false infeasibility detection. In particular, suppose an equation is declared infeasible if its (computed) least value exceeds its right-hand side by μ. If $\mu \geq \tau$, this happens in the example!

There are two ways we deal with this anomaly. First, we choose the *infeasibility tolerance*, μ, to be less than the *fix tolerance*, τ. This helps, but if we set μ too much less than τ, we miss detecting some legitimate infeasibility. In any case, we do not have a true safeguard by tolerances alone.

In the REDUCE algorithm we do one more thing, which exploits problem structure, and it works in this case. That is, suppose we decide to fix a variable whose (reduced) bounds define the interval [L, U] for the level of the variable. Instead of setting its level to the mid-point, we first test if either bound is an original one. If so, that is the fixed level chosen. If not, we see if the interval contains an integer. If so, that is the fixed level. Otherwise, as a last resort, the mid-point is chosen.

In ATESTOUT, which was created when you EXECUTEd ATEST after installation, this problem is run 25 iterations (the default). It fails to fix the levels of these two variables, but if you run REDUCE for 43 (or more) iterations, it correctly fixes their values at (2, 0). In this case it succeeds because each variable has one of its bounds remain unchanged, so they become the (correct) fixed levels.

In theory, this should always happen in such a case. If a variable must equal one of its bound values in every feasible solution, that bound should not change during the successive bounding algorithm. The purpose of presenting this anomaly is to illustrate that it is possible to obtain a false indication of infeasibility and that fixed values may not be precise. The method of fixing the variable, however, sometimes avoids the problem. In general, the ANALYZE procedures should be regarded as heuristic in the sense that the information and conclusions given may be incorrect. On the other hand, knowing this can lead to improved model management and better formulations that avoid such degeneracies.

7.3 Assisting model management.

In this section we describe various ways ANALYZE can be used to assist *model management* - that is, the ongoing maintenance, documentation and operation of the model. The use of rule files, particularly with the English translations, offer one immediate aid to operations, especially if the user constituency is unfamiliar with linear programming.

Syntax information, especially the SCHEMA command, can assist documentation not only of instances, but of the model, itself. One may, for example, compare the set and coefficient information in the schema with what should be there. This is one form of *verification*: determining that what is there is what is supposed to be there. The TALLY command can be used in the same way for verifying sets that comprise the domains of equations and activities.

The use of blocking, particularly to find embedded netforms and components, is a similar aid for verification and documentation. To some extent this can be automated and a utility written to compare the ANALYZE output with what is expected.

Another approach is to use ANALYZE during model formulation as an aid for subsequent model management. Analyzing prototypes can reveal *tension points* of the model - that is, portions that are critical for either diagnosing an unsuccessful run (like infeasible) or interpreting the levels and prices in an optimal solution. Such prototyping can also aid the construction of rule files to help analysts during the model's operations.

The BASIS and REDUCE commands can reveal redundancies in enough instances to offer guidance for model review and for building a *check file* of things to examine before a run is considered successful. To some extent a check file can be automatically created with some additional software that uses the results of ANALYZE. In some cases ANALYZE may reveal that the structure of the model, not the particular data, poses a potential problem.

One such case is when a submatrix is constructed from the sign pattern of nonzero coefficients, such as with the TRACE command. (The BASIS command's structural display can help here too.) For example, suppose a cycle is revealed - that is, a complete submatrix that cannot be rearranged to triangular form. After accounting for scales (in cases of non-unit coefficients), the total cost of this cycle cannot be negative (for minimization) if it is uncapacitated (i.e., no activities in the cycle have upper bounds). This can be written to a check file, so the data is checked for this before optimizing, perhaps only when the relevant database is changed. A negative cycle results in a diagnostic and a fatal error to prevent entering the optimizer. A (nearly) zero cycle may result in a nonfatal error, but with a warning message.

More sophisticated assistance for model management may require access to the ANALYZE source code to build an efficient utility, especially with automating the generation of a check file. This access is described in chapter 12.

As an extended part of the formulation responsibilities, the special files can serve both to assist analysis and to manage the model. For any particular instance, the sequence shown in figure 7-18 can be prepared for an ANALYZE session. By defining it to be the PROFILE execution file and by entering QUIT at the end, this may be regarded as a batch job.

```
READIN MATRIX name
SOLUTION filespec
BASIS
WRITEOUT
READIN SYNTAX
SCHEMA DISPLAY
SCHEMA SAVE
QUIT
```

Figure 7-18. A Batch Sequence to Prepare for an ANALYZE Session

The "batch" processing is to avoid time-consuming procedures for large models. The most time-consuming is reading the matrix file, which is the first command in the sequence. Then, the solution file is loaded. The BASIS specification is to prepare the basis for some of the commands, notably for the RATEOF command, which gives the rates of substitution between nonbasic and basic variables. The WRITEOUT command then writes a packed file, which now has the matrix and the solution; and, the basis is setup. This packed file will load quickly, which we shall discuss below.

Suppose a syntax file, *name*.SYN, was prepared. Then, after it is read into memory, the first SCHEMA specification forms the schema table (with the DISPLAY option). The second saves the schema for fast loading.

Now consider the preparation of an associated execution file, *name*.EXC, shown in figure 7-19.

```
SWITCH MSG
  _SETUP PACKED \LPDATA\name\
  _SETUP SYNTAX \LPDATA\name\
  _SETUP MATRIX *  .MPS
  READIN PACKED name
  READIN SYNTAX
  PRINT \LPDATA\name\name
  SCHEMA LOAD
$ a message can go here
  SCHEMA DISPLAY *
  INTERPRT STATS
  _DICTNRY MERGE \LPDATA\name\name
SWITCH MSG
RETURN QUIETLY
```

Figure 7-19. An Execution File to Begin an ANALYZE Session

The first line turns off the message switch merely to avoid the brief messages that are otherwise printed on the screen. (This is turned back on just before the RETURN specification.) Afterwards, the _SETUP command is used to set the prefixes and suffixes of the standard file types that are not in the default path. In this case, we change the prefixes of the PACKED and SYNTAX file types to be \LPDATA*name*, where *name* is where we have those files for this session. Then, _SETUP is used to change the suffix of the MATRIX file to be .MPS (instead of .MAT). This may be because other matrix files will be read, and other modeling systems and solvers produce and expect, respectively, the .MPS suffix.

The first specification of the READIN command loads the model instance, as it was written in the batch sequence (c.f., figure 7-18); the second reads the syntax file into memory. The PRINT gives the user an introduction that was prepared in \LPDATA*name**name*.DOC (we did not use _SETUP to change the PRINT prefix because the user may want to PRINT other files that are in the default path, such as those that came with ANALYZE). After the schema is loaded (which was previously saved), messages can be given to let the viewer know what the schema display means.

In this simple example, we end the introduction by using the INTERPRT command with the STATS rule file (which came with ANALYZE, but can be changed to suit the particular constituency). Finally, a special dictionary file, *name*.DIC, is merged with the ANALYZE dictionary (already resident unless prevented by not having ANALYZE.DIC in the default path, which may have been changed by the SETUP.EXC file). The MERGE could be replaced with a LOAD if the ANALYZE dictionary is inappropriate. Please note that a merge is not safeguarded against duplicate entries.

A more elaborate scheme to prepare for sessions with a batch file is possible. In particular, rule files can be prepared and applied from the execution file in certain cases, such as an infeasible instance. Some of the rule files can be bound specifically to this model and executed from a generic rule file. Figure 7-20, for example, illustrates a rule file that is executed by the user's specification. If the problem *name* matches one of those listed, a special rule file for that model is executed (but kept hidden from the user, as it does not appear in the options of the INTERPRT command). Otherwise, a generic message is printed.

```
* File:  INFEAS.RUL
* Function: Execute infeasibility rule file
*           for the particular problem (rule
*           suffix = .INF)
*
$IF %PROBLEM = name1 THEN INTERPRT name1.INF
$IF %PROBLEM = name2 THEN INTERPRT name2.INF
               •
               •
               •
$IF %PROBLEM = namen THEN INTERPRT namen.INF
*       PROBLEM IS NOT IN LIST
Sorry, there is no rule file for interpreting the
infeasibility of %PROBLEM .
$EXIT
```

Figure 7-20. A Rule File to Interpret Infeasibility for the Particular Problem

Creative use of the various files to support analysis is encouraged, and you may report such innovations (as well as problems) to the author (see cover for bitnet and snailmail addresses).

Chapter 8. Utilities

In any investigation of truth, three controlling motives may be present: first, that the truth in question may be found; then, that being found, it may be demonstrated; and that, once demonstrated, it may provide a means to distinguish truth from error.
Blaise Pascal

This chapter describes some utilities that may be useful, apart from ANALYZE. The HPAGER utility converts text files produced by ANALYZE to obey page ejection for an HP Laser Jet (or any printer with the same control character for page ejection). The LPRENAME utility converts LP file inputs into those recognized by ANALYZE. Its primary function is to permit renaming of rows and columns to allow longer names than the 8-character limit imposed by a standard matrix file. The FORTREF utility (which comes only with the full version) reads FORTRAN source files and forms a *program network*. Its main purpose is to give a sorted list of subroutines, the files that contain them, their predecessors (i.e., subroutines that call them), and their successors (i.e., subroutines they call). Other uses are described, including overlay experimentation.

The utilities are not automatically extracted with the ANALYZE installation. To do so, change to the directory you want the utility to reside. Then, put the UTILITIES disk into drive a; and, at the DOS prompt enter:

> A:PKXARC A:*file* INSTALL.BAT

where *file* is one of: HPAGER, LPRNDATA, or FORTREF. Then, at the DOS prompt enter:

> INSTALL [*drive*]

The default drive is a, so enter INSTALL B if your floppy drive is b. This will automatically install *utility* onto your hard disk, after which you may delete INSTALL.BAT from your hard disk.

It is advisable to install each utility in its own path to avoid file name conflicts.

8.1 HPAGER

If you have an HP Laser Jet, or any printer that uses the same control character for page ejection, you can use the HPAGER utility to replace those 1H1's with the page ejection character. It is necessary to give a different file name to the output of HPAGER when it prompts you. For example, suppose you obtain

ANALYZE.DOC from EXEC DOCANAL. Tell HPAGER that the output file is ANALYZE.HP or ANALYZE.PRT (or some name other than ANALYZE.DOC).

Example:

```
> hpager
NAME OF INPUT FILE: hpager.doc

HEADER:  this is a header

NAME OF OUTPUT FILE: hpager.prt

OK, CONVERTING hpager.doc                     TO hpager.prt
EOF REACHED AT LINE=          18...        0 pages
NAME OF INPUT FILE:
```

After entering `hpager` at the DOS prompt, the uppercase prompts should appear on your screen. The lower cases comprise the responses (except in the confirmation message). After it finishes with the End-Of-File (EOF) message, HPAGER starts over in case you want to convert another file. Press Enter to return to DOS. You can examine HPAGER.DOC, which came with your utility, and you will see it is the same as HPAGER.PRT because it contains no page ejection. A more interesting example is ANALYZE.DOC, which you can create with EXEC DOCANAL in ANALYZE.

Note that HPAGER says there are 0 pages. This is because the page numbering starts at 0, allowing for a cover page, as with the reference manual generated by DOCANAL.EXC. If a header is specified, the pages are numbered, starting with page 1.

8.2 LPRENAME

LPRENAME is designed to permit renaming of rows and columns. This provides an interface with some modeling languages, like GAMS, that generate innocuous names, like ROW001,... and COL001,... that contain no syntax. These names can be automatically converted to describe the classes and domains that may require greater length than the 8-character standard in a matrix file.

The introduction and on-line help give examples of how to use LPRENAME. For general interfacing with other systems, a new file type is defined, called simply an *LP file*. This contains matrix and solution values plus names that can be up to 16 characters.

Overall, the LPRENAME I/O is shown in figure 8-1. A necessary input is a matrix file, a packed file, or a LP file. Optionally, one can read a solution file into memory. The READIN and SOLUTION commands are the same as in ANALYZE. Similarly, the WRITEOUT command produces a packed file, and it is also the same as in ANALYZE.

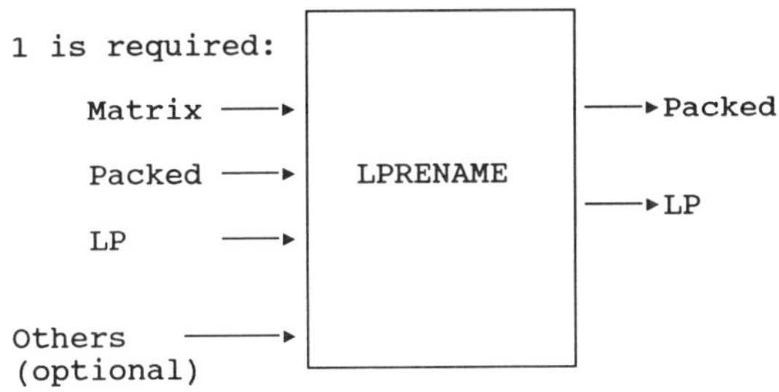

Figure 8-1. I/O Files for LPRENAME

If you use an optimizer different from the ones recognized by ANALYZE, you can write your own solution file that conforms to one of the recognized formats, then use the RENAME command to enter the correct name of the solver. The simplest format to explain is for the OSL solver, as follows.

First, the file is read as a standard ASCII file, where blank lines are ignored and an asterisk in column denotes a comment. The first (non-comment) line of the OSL solution file contains the problem name, sense of optimization (MIN or MAX), the overall solution status (OPTIMAL, INFEASIBLE, UNBOUNDED, FEASIBLE, or Unknown), the names of the RHS, RANGE and BOUND sets. The three names are compared with what was read from the matrix file to verify that the solution file contains the same LP. Note that all names are case-sensitive, so a keyword like MIN must be all caps.

Following the header line is the rows section, followed by the columns section. Each of these sections have the same format: name status level price. All lines are free field. There must be one line per row in the rows section, and one line per column in the columns section. For an example, see WOODNET.OSL, which came with the ANALYZE data.

One of the optional input files is a *rename file*. Its purpose is to rename rows and/or columns by specifying the oldname, then the new name. First, however, the name length and other specifications (like the problem name) may need to be changed from their current setting. The RENAME command is used for setting the *base* of the LP.

As you can see with the on-line help, the RENAME command syntax is as follows.

```
RENAME [ LENGTH = number
       | PROBLEM = name
       | RHS = name
       | BOUNDS = name
       | RANGE = name
       | SOLVER = name
       | USING [filespec] [//{ROW|COLUMN}]
```

Each option, except USING, is a base setting. The LENGTH is the length of the row and column names. It can be any integer between 1 and 16. The current setting is at least as great as the current name lengths, but you need to increase it if the new names are longer. If the new names are shorter, first perform RENAME USING, then reduce the length before writing the output file.

The USING option reads a rename file (use _SETUP to change prefix and suffix of this file type). If the ROW or COLUMN parameter is specified, each line of the file is prefixed with it. Figure 8-2 illustrates this.

```
ROW   R1   SMO
ROW   R2   SOK
ROW   R3   DMO
ROW   R4   DOK
COL   C1   TMO
COL   C2   TOK
```

(a) RENAME USING

```
R3  DMO
R2  SOK
R1  SMO
R4  DOK
```

(b) RENAME USING //ROW

Figure 8-2. Using a Rename File

In figure 8-2 (a), each line specifies whether it is a row or column to be renamed. These can come in any order. In (b), only the rows are to be renamed, and the file does not contain the ROW prefix. This is inserted by the specification of //ROW in the RENAME USING command.

The following sequence reads a matrix and solution file, called MYPROB.MAT and MYPROB.MNS, respectively, into memory. Then, the name length is re-set to 12 characters before renaming the rows and columns. This is done with 2 rename files: MYROWS.REN and MYCOLS.REN. Finally, the packed file is written, which can then be read in by ANALYZE.

```
REA M MYPROB
SOL MI
REN L=12
REN U MYROWS //ROW
REN U MYCOLS //COL
WRITE
```

This sequence could be put into an execution file. It can be made to seem like a batch job if the execution file is defined as the PROFILE and it has a QUIT specification at the end. Then, LPRENAME immediately executes this sequence and quits.

An LP file is an ASCII file that contains all matrix and solution information. It can be saved and loaded with the LPRENAME commands, SAVE and LOAD, respectively. Figure 8-3 shows the structure of an LP file.

`key = name or value`
`.`
`.`
`.`
`ENDBASE`
	Base

`name nonzeroes lower_bound upper_bound`
`status level price`
`.`
`.`
`.`
	Rows segment
	(2 lines @)

`name nonzeroes lower_bound upper_bound`
`status level price`
`.`
`.`
`.`
	Columns segment
	(2 lines @)

`row_name column_name value`
`.`
`.`
`.`
`ENDATA`
	Nonzeroes segment
	(1 line @)

Figure 8-3. Structure of LP file (all free field)

The Base segment announces problem information with keywords, shown in figure 8-4. The only required specifications are the NAME, ROWS, COLUMNS and NONZEROS. The ENDBASE line is required.

Base element	Meaning
NAME	Problem name
OBJECTIV	Objective row name
LENGTH	Row and column name lengths (1 to 16)
RHS	Right-hand side set
BOUNDS	Bounds set
RANGE	Range set
SOLVER	Solver (e.g., OB1, MINOS)
STATUS	Solution status: {OPTIMAL, MINIMUM, MAXIMUM, FEASIBLE, INFEASIBLE, UNBOUNDED}
OPTIMAL	Sense of optimization: {MINIMIZE, MAXIMIZE}
ROWS	number of rows
COLUMNS	number of columns
NONZEROS	number of nonzero coefficients

Figure 8-4. Base Elements

The row and column segments have 2 lines per row and column, respectively. Each name is 1 to L characters, where L is the LENGTH value. If the base segment did not specify a length, the current setting is used, which may have been specified by the RENAME command before loading the LP file.

The nonzeroes of each row and column is the number of nonzeroes in it, exclusive of right-hand sides and bounds. The sum of the row nonzeroes should equal the sum of the column nonzeroes, and this should equal the NONZEROS specification in the base segment. The lower and upper bounds pertain to the row and column levels, y and x, respectively, where $y = Ax$. Infinite bounds are indicated by an asterisk (*).

Line 2 of each row and column begins with the solution status. The recognized codes are: B,I,L and U, where

B .... Basic (and feasible);
I.......... Infeasible (and basic);
L......... Nonbasic at lower bound;
U........ Nonbasic at upper bound.

Then, the level and price, associated with the resident solution, are given in F or E formats.

After executing ATEST, some LP files are generated, which you can examine. On-line help is given in LPFILE.DOC.

The edit commands, DELETE, FILL, and REPLACE, provide renaming operations that may be necessary if linking with a modeling language. For example, DELETE [deletes the left bracket from row and column names, which may have been put into the input file by AMPL. The DELETE command left-justifies the resulting string - that is, the row or column name. To keep the string as is, use the REPLACE command. For example, REPLACE [_ replaces the left bracket with an underscore; and, REPLACE [replaces the left bracket with a period, which is the default replacement character. The FILL command is a special replacement of blanks with some ASCII character. For example, FILL _ fills embedded blanks with the underscore; and, FILL fills them with periods, which is the default fill character.

On-line help is available for each command, and QUICKREF.DOC contains a quick reference to their functions. A reference manual can be obtained by entering EXEC DOCLPREN.

8.3 FORTREF

To help manage the 300 subroutines (approximately) in separate modules, an interactive program, called FORTREF, is designed to read the source files and construct a program network. The on-line help should be sufficient to see what it does, and you can look at the execution files (*.EXC) to see how to use FORTREF to obtain the cross-reference information in ANALYZE.REF. (FORTREF is included only if you have the full version, and it is on the utility disk.)

FORTREF has commands to experiment with different overlays. See the OVERLAY command by specifying HELP OVERLAY (or simply H OV), which shows some examples. The rule is that if subroutine A calls subroutine B or visa-versa, the files containing A and B must either be in the same overlay or one of them must be in the main segment. The FILES command gives this information. You can also use the SHOW command to see the predecessors and successors of a particular set of subroutines, specified by a name mask.

FORTREF also performs topological sorting with the SORT command, and it prints a loop if one is present (such is the case in ANALYZE, except for CMS where re-entrance is not allowed). Look at the execution files that came with the system, notably FORTREF.EXC and ANALYZE.EXC. (Adjust these for the subdirectory into which you loaded the source codes and link file.) Execute them and note the output from each command.

The CHECK command checks that all subroutines have been defined and that there is exactly one main program. If these conditions hold, CHECK proceeds to check the connectivity of the program network by finding a path from the main program (called .MAIN) to each subroutine. It is not fatal if some routines are not connected. In fact, the random value routines in the FLIP library are not used by ANALYZE, so they will appear as disconnected from the main program.

Chapter 9. Algorithms and Heuristics

There are reports that many executives make their decisions by flipping a coin or by throwing darts, etc. It is also rumored that some college professors prepare their grades on such a basis. Sometimes it is important to make a completely "unbiased" decision; this ability is occasionally useful in computer algorithms, for example in situations where a fixed decision made each time would cause the algorithm to run more slowly.

Donald E. Knuth

This chapter describes fundamental algorithms and heuristics used by ANALYZE for some of its commands. The purpose of this presentation is to deepen one's understanding of the results obtained from syntax-based explanations (EXPLAIN and SYNTAX commands) and from the TRACE, BASIS, RATEOF and REDUCE commands.

9.1 Discourse formation.

Here we describe the syntactic translations of rows and columns from the information provided by a syntax file (see §6.6). There are two aspects: specific rows and columns translated by the EXPLAIN command, and class translations by the SYNTAX command (which also applies from within a rule file).

The syntax of a row or column defines a *syntax map* from its name to a sentence. The form of a syntax map for a row or column is a *template* of literals and *set references*. Each set reference is designated by a special character (&), followed by the k-character code of the set and digit(s) that determine the substring position in the name to be translated (on-line documentation of the syntax file structure is in FILESYN.DOC) . The value of k determines the length of the field and the digit(s) determine its beginning. Typically, k=1 or k=2. In the first case, the set has a 1-character code, and each of its members is a 1-character code. In the second case, the set has a 2-character code, and each of its members is a 2-character code.

For example, consider the following syntax map.

```
P  produces &MT:2 in &LO:4 by &M:6
```

This says each column whose name begins with P translates according to the following steps.

1. Begin string with literal portion of `P` (i.e., until the first set reference). In this case the string becomes:

   ```
   produces
   ```

   and we reach the set reference `&MT:2`.

2. Look up the set and use the name substring determined by the digits of the set reference as a key into the set's members from which the translation is substituted into the main string. After substituting the set member's translation, return to step 1.

In our example, we look up the set MT, which we may suppose is a material, and we use characters 2 and 3 from the column name as a key. In particular, suppose we have a column named POLSW2. Then, we look up OL in table MT, which suppose has the entry:

```
MT material            ◄—— set
      OL crude oil     ◄—— member
```

Our string then becomes:

```
produces crude oil
```

Returning to step 1, we add the literal (in) and reach the next set reference (&LO:4). There must be a set named LO, which we may suppose means location. In particular, suppose we have:

```
LO location            ◄—— set
      SW Southwest     ◄—— member
```

Then, the translation string becomes:

```
produces crude oil in Southwest
```

Again, returning to step 1, we reach the last set reference (&M:6), which refers to set M, say mode of production, and determines that the key is the 6-th character of the name. In particular, suppose we have:

```
M mode                          ◄—— set
      2 secondary drilling      ◄—— member
```

Then, the final translation is:

```
produces crude oil in Southwest by secondary drilling
```

The EXPLAIN command of ANALYZE uses this in the *template*:

Column <*name*> &ct .

where Column is literal, and <*name*> is the name of the column that is substituted. Finally, &ct is the *column translation rule* using the syntax map, just described. The response for our example is thus:

```
Column POLSW2 produces crude oil in Southwest by secondary drilling.
```

Each column and row is translated in the same way using the syntax described by the syntax file. The SYNTAX command also uses the syntax maps, but with "`some`" substituted for specific set look up. Here is the algorithm:

1. Begin the translation with: `A column that begins with` <*class*>, substituting for <*class*> the character(s) of the column class (rows proceed similarly). Then, proceed to substitute the literals in the syntax map until the first set reference. In our example, the first set reference is &MT:2, which follows the literal `produces`, so we have:

   ```
   A column that begins with P produces
   ```

2. Continue the main string by adding `some` (literally), followed by the translation of the set. In our example, this becomes:

   ```
   A column that begins with P produces some material
   ```

Returning to 1, we continue to form a *class translation*. Our example becomes:

```
A column that begins with P produces some material in some location
    by some mode.
```

9.2 Path tracing.

Here we describe the heuristic used by the TRACE command. The purpose of the trace procedure is to determine a substructure that helps to explain a flow or price. It begins with a submatrix and adds rows and columns according to the criteria described below until a *complete* submatrix is obtained. A submatrix is complete if the following conditions hold for each row in the submatrix.

If the row is G-type (i.e., $Ax \geq b$),	it must contain a positive coefficient.
If the row is L-type (i.e., $Ax \leq b$),	it must contain a negative coefficient.
If the row is E-type (i.e., $Ax = b$),	it must contain both signs.

Free rows play no role, and ranged rows are treated as their binding inequality dictates, according to the resident solution.

For example, the following submatrix is complete.

```
+  -         =  0
   +  -      =  +
+     +     >=  +
   -  -     <=  -
```

The idea is that it is possible to increase marginally the levels of the activities in the submatrix and remain feasible.

It is the *possible* perturbation that we seek, ignoring the magnitudes of the coefficients and right-hand sides. Only sign patterns determine the final submatrix of a trace.

If the submatrix is not complete, the columns are interrogated. To be a candidate to enter the submatrix, a column must *cover* at least one *uncovered* row. A row is uncovered if it is in the submatrix but violates its condition for completeness. Such rows enter the submatrix either by the initial specification or by having a nonzero in a column that is added. A column covers an uncovered row if its coefficient supplies the needed sign.

Among candidates that cover some uncovered row, there are two figures of merit used. The first is the *net cover* of the column. This equals the number of rows already in the submatrix that become covered upon adding this column (which must be at least 1) minus the number of uncovered rows that would enter the submatrix. This net cover is allowed to be negative, and the greatest net cover is preferred. The second figure of merit is the status, relative to the resident solution, called the *rank* of the column. Since we are seeking an explanation of perturbation, we use the following rank definitions.

Rank		
Rank	$= 1$ if $L > X$ or $X > U$	(i.e., Status = Infeasible)
	$= 2$ if $L < X < U$	(i.e., Basic and non-degenerate)
	$= 3$ if $L = X < U$ and $D \le 0$	
	$= 4$ if $L < X = U$ and $D \ge 0$	
	$= 5$ else.	

The rationale for this ranking is that if the LP is infeasible, the most preferred column to add to the submatrix is one that is infeasible. Otherwise, if LP is feasible, we prefer room for the activity level (X) to vary marginally. The most usual use of the TRACE command is with a resident optimal solution, in which case there are no rank 1 columns.

For rank 2 the activity must be basic (assuming the resident solution is basic, such as generated by a variant of the simplex method), in which case $D=0$. For rank 3, if the resident solution is optimal, we have $D \ge 0$ (for min), so rank 3 really implies $D=0$. This is a degenerate situation: either the activity is basic in the primal at its bound (L), or its dual complement (D) is. Similar comments apply to rank 4. Rank 5 includes fixed variables ($L=U$) and conditions under which the solution does not satisfy the primal-dual optimality conditions.

The algorithm first initializes cover properties of the rows in the initial submatrix and adds rows with coefficients in the initial columns. Then, if all rows are covered, the algorithm stops. Otherwise, the general iterations begin.

At a general iteration, at least one row is uncovered, so the admissible columns are interrogated. A column is admissible if it is not already in the submatrix. If no column covers an uncovered row, the algorithm stops (with a message). Otherwise, when a candidate column is found (i.e., it covers an uncovered row), its net cover is computed. If this is worse than the current best, it is dismissed; otherwise, its rank is computed. If its rank is 1 or 2, the column is immediately added, and the iteration is completed by adding its uncovered rows to the submatrix. If its net cover is strictly better than the current best or its rank is strictly better than the current best, the column becomes the current best; otherwise, it is dismissed.

The iterations continue until a complete path is obtained or uncovered rows cannot be covered. There is no backtracking, so an early decision to accept a column could miss an alternative construction.

This algorithm is sometimes called a *path trace*, because it obtains a path in the usual sense if the LP is a network model. It is not necessary, however, for the LP to be an ordinary network (it is sometimes insightful to think of the LP as a *process network*, in which case a path is that of a hypergraph).

9.3 Basis rearrangement.

Here we describe the heuristic used by the BASIS command when setting the factors to support transformations. The general idea is to permute the rows and columns of the basis to make the rearranged matrix lower triangular if possible. There are two running pointers, *front* and *rear*, which point, respectively, to the first and last matrix positions for which no pivot assignment has been made.

The algorithm begins by assigning all logicals (i.e., row variables in the basis) to the rear because they are column singletons. After marking their associated rows (to exclude pivot assignments to those rows), triangularization is attempted by assigning row singletons to the front. This is recursive, so each assignment can create a row singleton in the reduced submatrix. If this fails to assign all basic variables, this means the basis cannot be triangularized. Some columns must be *spike* - that is, have a nonzero above the main diagonal of the rearranged basis, no matter how the rearrangement is chosen.

If the forward triangularization terminates with an incomplete assignment, backward triangularization is executed. This is the recursive assignment of column singletons to the rear. The result is a partial assignment to two segments: *forward triangle* and *backward triangle*. The middle portion is called the *kernel* of the basis.

The basis statistics, given by the BASIS command, describes how many basic activities are in the forward triangle, the backward triangle and the kernel. The following algorithms explain what these sections mean.

Basis Agenda Algorithm

1. Execute FTRI (forward triangularization). If all columns are assigned, exit.

2. Execute BTRI (backward triangularization).

3. Select the best assignment in the kernel until all columns are assigned.

4. Assign each spike as early as possible and compute α-vectors.

Routine FTRI (*Forward Triangularization*). Among unassigned rows, find one with min count. If this count is 1, assign the row and its (only) unassigned column to the front and repeat; otherwise, exit.

Routine BTRI (*Backward Triangularization*). Among unassigned columns, find one with min count. If this count is 1, assign the column and its (only) unassigned row to the rear and repeat; otherwise, exit.

Kernel Assignment (simplest variation). Evaluate the assignment of a row of min count to the front with one of its columns (this forces its other intersecting columns to be spike) to be its count minus 1. Evaluate the assignment of a column of min count to the rear (which is spike) to be the new min row count (after decrementing the row counts for those rows having a nonzero in this column). Each evaluation is an estimate of spikes, which we seek to minimize, so choose the assignment (front or rear, respectively) of lesser evaluation.

Assignments

If (ROW,COL) is assigned to the front, the remaining unassigned rows intersecting COL have their counts decremented by 1. If there are any other unassigned columns intersecting ROW, they become spikes. For each of these spikes, their intersecting, unassigned rows have their counts decremented.

> If (ROW,COL) is assigned to the rear, the unassigned rows (columns) intersecting COL (ROW) have their counts decremented.

Each assignment makes it possible to resume triangularization for the remaining submatrix. After the kernel assignment is completed, we return either to FTRI (if the assignment is to the front) or to BTRI (if the assignment is to the rear).

Consider the example shown in Figure 9-1. This 8×8 basis has two row variables (i.e., logicals), R7 and R8, and 6 column variables (i.e., structural activities). The nonzeroes are indicated by the asterisk (*). We begin the basis agenda algorithm by assigning the row variables to the rear, which is their position shown. The markers and counts are initialized for the remaining 6×6 submatrix.

	C1	C2	C3	C4	C5	C6	R7	R8
1	*	*		*				
2	*		*		*			
3		*	*		*			
4	*							
5				*				
6	*		*	*	*	*		
7		*			*		*	
8		*				*		*

Figure 9-1. An Example Basis

Figure 9-2 shows the initial information. The column counts are exclusive of the last two rows, which are already assigned to their row variables, R7 and R8, respectively. The *count* is the number of nonzeroes in unassigned rows/columns. (The actual bookkeeping in the implementation is more involved; it is described in the source code listings.)

	C1	C2	C3	C4	C5	C6	R7	R8	Count
1	*	*		*					3
2	*		*		*				3
3		*	*		*				3
4	*								1
5				*					1
6	*		*	*	*	*			5
7		*			*		*		na
8		*				*		*	na
Count	4	2	3	3	3	1	na	na	

Figure 9-2. Initializing Row and Column Counts

Upon entering Routine FTRI we find row 4 is a singleton, so it is assigned to the front with column C1. Then, row 5 is assigned to the (new) front with column C4. These assignments reduce the count of row 1 to 1, so the second pass assigns row 1 to the front with column C2. The final result of FTRI is shown in Figure 9-3.

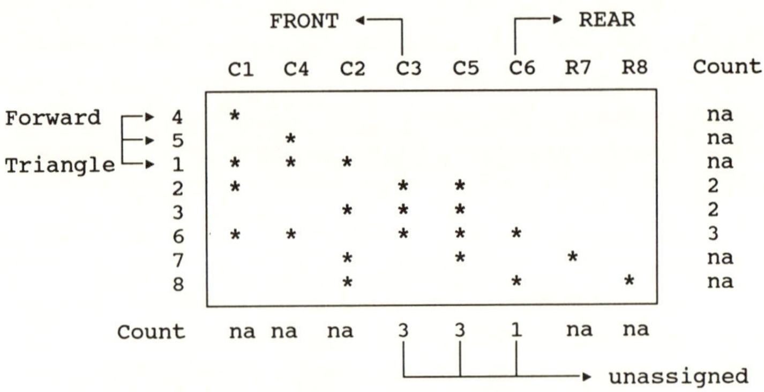

Figure 9-3. Result of Forward Triangularization

The forward triangle, found by Routine FTRI, has columns C1, C4 and C2 pivoting, respectively, on rows 4, 5 and 1. The resulting counts are for the remaining 3×3 submatrix. Next, Routine BTRI is entered, and the column singleton C6 is assigned to the rear with pivot row 6. The result is shown in Figure 9-4, with the three segments partitioned: forward triangle, kernel and backward triangle.

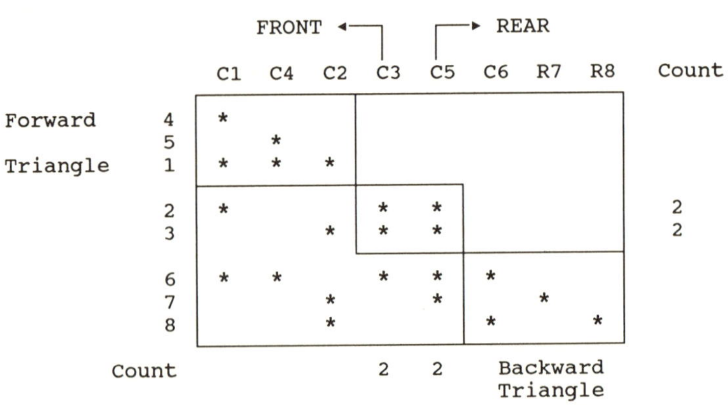

Figure 9-4. Result of Backward Triangularization (after FTRI and BTRI)

In summary, the objective of the basis agenda algorithm is to find as nearly a triangular rearrangement of the basis matrix as possible. All network models completely triangularize, which saves both space and time, especially when computing the rates of substitution. Most large LP's have large embedded networks, so there are few disturbances to this triangularity. Still, there are some. Here the essential features of the basic agenda algorithm was described; details of the implementation are given in the source code listings in file GBASIS.FOR. There are many variations possible, so computational bottlenecks should be reported to improve later versions.

Once the basis agenda is determined, except possibly for the pivot row assignments to the spike columns, the α-vectors are computed. For a non-spike column, its α-vector is its original column. In that case it is simply copied to the special array, putting the pivot row first (and tagged to separate the α-vectors).

Each spike is recorded with its earliest insertion possibility. For example, if a column's assignment to the front caused column j to be spike, its earliest insertion is immediately after the assigned column. Its α-vector is computed, and it is assigned to its earliest position if it has a nonzero α coordinate (greater than some tolerance) in an unassigned row. Otherwise, it is stacked for later consideration. After all early insertions are determined, new α-vectors are computed for the stacked spikes. This may result in more stacking. The factorization is complete once all spikes have entered the α-vector region. Failure to obtain an acceptable pivot row for a spike results in an error message, indicating that the basis is too close to being singular for reliable computations. In that case the basis factorization is aborted, and the associated commands (like RATEOF) cannot be used.

9.4 Rates of substitution.

Here we explain the FTRAN and BTRAN algorithms used to compute the rates of substitution for the RATEOF command. Using the same notation as in the previous section, the two systems of equations are: $B\alpha = a$ and $\pi B = b$. Typically, a is a column of A, so α is its transform, $B^{-1}a$; and, b is a cost vector, say c_B, for the basic variables, in which case π is the dual price vector, $c_B B^{-1}$. More generally, these transforms are used to determine rates of change for rather arbitrary impulse vectors that could be data or variable.

The explicit inverse, B^{-1}, is not computed to determine α and π. Instead, the basis (B) is factored into elementary product form, as described in the previous section. Let

$$B = E_1 \ldots E_k,$$

where each E_i is an *elementary matrix* – that is, it differs from the identity matrix in exactly one column, say p_i, which is called its *pivot position* (row or column, as they are the same position).

Associated with E_i is its α-vector, say $\alpha_i=(\alpha_{i1}, ..., \alpha_{im})$, where $\alpha_{ip} \neq 0$ for $p=p_i$:

$$E_i = \begin{bmatrix} & \alpha_{i1} & \\ I & \vdots & 0 \\ & \alpha_{ip} & \\ 0 & \vdots & I \\ & \alpha_{im} & \end{bmatrix} \begin{matrix} \\ \\ \longleftarrow \text{pivot row} \\ \\ \\ \end{matrix}$$

(the column $\alpha_{i1}, \ldots, \alpha_{im}$ is the α–vector)

Using this notation, FTRAN(a) performs the following computations. First, the vector a is copied to a double precision vector, say z^0. The i-th iteration operates on z^i to obtain z^{i+1}, as follows (for $i=0,...,k-1$).

1. For $p=p_i$, compute $z_p^{i+1} = z_p^i/\alpha_{ip}$.
2. For $j \neq p_i$, compute $z_j^{i+1} = z_j^i - z_p^{i+1}\alpha_{ij}$.

Note that $z^{i+1}=z^i$ if $z_p^i=0$; otherwise, the nonzero α values affect the z-vector. Once z^k is computed, it is copied to the original a-vector (single precision) and returned as the FTRAN value. That is, the returned vector is the computed solution to $B\alpha=a$.

The BTRAN(b) routine performs the following computations. First, the vector b is copied to a double precision vector, say z^k. The i-th iteration operates on z^i to obtain z^{i-1}, as follows (for $i=k,...,1$).

1. Initialize $\sigma = 0$ (where σ is double precision).
2. For $j \neq p_i$, accumulate $\sigma \leftarrow \sigma - z_j^i \alpha_{ij}$.
3. For $p=p_i$, set $z_p^{i-1} = (z_p^i-\sigma)/\alpha_{ip}$.

Unlike FTRAN, one iteration of BTRAN changes only the pivot position value, z_p^{i-1}. Once z^0 is computed, it is copied to the original b-vector (single precision) and returned as the BTRAN value. That is, the returned vector is the computed solution to $\pi B=b$.

9.5 Redundancy test.

Here we explain the redundancy test employed by the BASIS REDUND command specification. Subroutine GBARNG (in file GBRANGE.FOR) obtains ranges of the basic variables. Using the resident basis, the basic equations are of the form: $u = Mv$, where u is the m-vector of basic variables, say $[x_B, y_B]^T$, and v is the n-vector of nonbasic variables, say $[y_N, x_N]^T$. Then, $M = B^* N^*$, where B^* is the (full) basis inverse:

$$B^* = \begin{bmatrix} -B & 0 \\ -R & I \end{bmatrix}^{-1}$$

→ unit columns for y_B (second column)

→ -basic columns from A (first column)

and N^* is the (full) nonbasic matrix:

$$N^* = \begin{bmatrix} -I & N \\ 0 & Q \end{bmatrix}$$

→ nonbasic columns from A (second column)

→ -unit columns for y_N (first column)

Note we changed notation because now B is a submatrix in the full basis, which contains row variables. The dimension of B is the number of active rows (y's that are nonbasic at a bound value), which equals the number of structural basic activities that comprise the columns of B.

The (i, j) coefficient of M may be computed in either of 2 ways. First, we could FTRAN(a), where a is any column from N^*, say the j-th. Then, after the FTRAN, the i-th coordinate of a is M_{ij}. Second, we could BTRAN(b), where $b = e_i$ is the i-th unit vector upon entering BTRAN. Afterwards, b is the i-th row of B^*. The inner product (b,a) is M_{ij} if a is the j-th column of N^*.

In computing the range of a basic variable, its pivot row is BTRANned. Then, looping over nonbasic variables, the inner product is computed to obtain its rate of substitution, M_{ij}.

The nonbasic bounds determine the range of this term of the i-th basic equation. Unlike the RATEOF command support, the ranging pre-processes the nonbasic rows to reduce their original bounds for the subsequent range computation. In particular, an L-row has an infinite lower bound ($-\infty \le y = Ax \le U$), which may be made finite by scanning the row's range with the original matrix coefficients and the x-bounds.

9.6 Successive bound reduction.

Here we explain the successive bounding algorithm used by the REDUCE command. The bounds on primal quantities and dual prices are successively reduced until either an infeasibility is detected or no further reduction is found. Table 9-1 gives the notation we shall use.

Array	Length	Meaning
XL	NCOLS	lower bounds for column levels (X)
XU	NCOLS	upper bounds for column levels (X)
YL	NROWS	lower bounds for row levels (Y=AX)
YU	NROWS	upper bounds for row levels (Y=AX)
YMIN	NROWS	myopic row min: $\min[y_i = A_i x:\ YL(i) \le y_i \le YU(i),\ XL \le x \le XU]$
YMAX	NROWS	myopic row max: $\max[y_i = A_i x:\ YL(i) \le y_i \le YU(i),\ XL \le x \le XU]$
PL	NROWS	lower bounds for row prices (P)
PU	NROWS	upper bounds for row prices (P)
DMIN	scalar	myopic min: $\min[d_j = c_j - \pi A_j:\ PL \le \pi \le PU,\ d_j \ge 0 \text{ if } XL(j) = -\infty,\ d_j \le 0 \text{ if } XU(j) = \infty]$
DMAX	scalar	myopic max: $\max[d_j = c_j - \pi A_j:\ PL \le \pi \le PU,\ d_j \ge 0 \text{ if } XL(j) = -\infty,\ d_j \le 0 \text{ if } XU(j) = \infty]$

Table 9-1. Notation for REDUCE Arrays

Initially, XL, XU, YL, and YU are set to their given bound values (infinity included). The price range PL, PU is initialized according to row type:

```
PL(i) = -∞ if YU(i) < ∞;  else, 0.
PU(i) =  ∞ if YL(i) >-∞;  else, 0.
```

At a general iteration, YMIN and YMAX are initialized to 0 upon beginning the column loop. The column loop accumulates YMIN and YMAX; and, it tests for dual infeasibility and forcing prices.

For each row (in the submatrix), the extreme column levels, XL or XU, are determined myopically according to whether the nonzero is positive or negative. At the end of the nonzero interrogation, DMIN and DMAX are set. If the column is already fixed (i.e., XL=XU), there is no further action beyond updating YMIN and YMAX. If some row was determined forcing on the last iteration, the column becomes fixed by setting XL = XU = extreme value. Otherwise, the dual test is performed.

The dual test is as follows. If DMAX is less than the required minimum reduced cost (0 if XU is infinite), we have dual infeasibility. Similarly, if DMIN is greater than the required maximum reduced cost (0 if XL is infinite), we have dual infeasibility. In either of these cases the algorithm terminates with dual infeasibility detection. Otherwise, DMAX and DMIN are compared to these limits. If either equals its threshold value (within tolerance), this forces the dual prices of the rows of the nonzeroes to their extreme values - that is, PL and PU become equal to the forced dual price, which might affect subsequent iterations for other columns.

Once the column loop is completed, the rows are interrogated to see if YMIN or YMAX yields a row that is primal infeasible, forcing or redundant. A row (i) is infeasible if YMAX(i) < YL(i) or YMIN(i) > YU(i). A row is forcing if either YMAX(i)=YL(i) or YMIN(i)=YU(i). A row's inequality, $y_i \geq$ YL(i), is redundant if YMIN(i) ≥ YL(i) and YL(i) is finite; and, $y_i \leq$ YU(i) is redundant if YMAX(i) ≤ YU(i) and YU(i) is finite. In such cases a price bound is tightened - that is, PU(i)=0 in the first case, and PL(i)=0 in the second case.

Iterations continue as long as some reduction occurs and we have not gone beyond the maximum number of iterations; otherwise, successive bound reduction terminates. Later, further tests will be implemented, and this document will be revised accordingly.

9.7 Logical implications test.

Here we explain the logical implications test used by the REDUCE command when //LOGICAL is specified. Let x_j denote the level of a binary variable. An *implication* has the form: IF x_j = v THEN x_k = v′, where v and v′ are either 0 or 1, and x_k is another binary variable ($k \neq j$).

The test performed for each (non-fixed) binary variable is to compute the range of a row variable, say y_i, with x_j at 0, then at 1. As in the previous section, let YMIN and YMAX be the range vectors, so every feasible value of x results in YMIN ≤ y ≤ YMAX, where y = Ax.

Let YMIN0, YMAX0 be the range of y when we set x_j=0; and, let YMIN1, YMAX1 be the range of y when we set x_j=1. Then, we have:

If $A_{ij} > 0$,

```
YMINO(i) = YMIN(i)
YMIN1(i) = YMIN(i) + Aij
YMAXO(i) = YMAX(i) - Aij
YMAX1(i) = YMAX(i)
```

If $A_{ij} < 0$,

```
YMINO(i) = YMIN(i) - Aij
YMIN1(i) = YMIN(i)
YMAXO(i) = YMAX(i)
YMAX1(i) = YMAX(i) + Aij
```

For example, suppose $A_{ij} > 0$ and we want the new value of YMIN(i) when x_j=0 and when x_j=1 - that is, YMIN0(i) and YMIN1(i), respectively. When YMIN(i) was computed, the value x_j=0 was used because $A_{ij} > 0$. There is, therefore, no adjustment needed to obtain YMIN0(i). It is necessary, however, to add A_{ij} when considering x_j=1. The other adjustments are similar, adding or subtracting A_{ij} when it is necessary to adjust YMIN and YMAX.

The test, for each nonzero A_{ij}, is now similar to the bound reduction test for forcing extreme levels. That is, suppose YL ≤ y ≤ YU are the limit constraints on the y-vector. Then, we apply the following rules.

If YMIN0(i) = YU(i), all variables in the i-th equation are forced to their extremal bounds for which y=YMIN; in particular, all other binary variables in the i-th equation are forced (by implication of x_j=0) to be either 0 or 1.

If YMIN1(i) = YU(i), all variables in the i-th equation are forced to their extremal bounds for which y=YMIN; in particular, all other binary variables in the i-th equation are forced (by implication of x_j=1) to be either 0 or 1.

If YMAX0(i) = YL(i), all variables in the i-th equation are forced to their extremal bounds for which y=YMAX; in particular, all other binary variables in the i-th equation are forced (by implication of x_j=0) to be either 0 or 1.

If YMAX1(i) = YL(i), all variables in the i-th equation are forced to their extremal bounds for which y=YMAX; in particular, all other binary variables in the i-th equation are forced (by implication of x_j=1) to be either 0 or 1.

Note that when logical testing begins, YMIN0 ≤ YU, YMIN1 ≤ YU, YMAX0 ≥ YL and YMAX1 ≥ YL; otherwise, the successive bound reduction would have detected infeasibility.

To illustrate the application of these rules, consider the example that came with ANALYZE (matrix file is LGLTEST.MAT). The LP has the following constraints.

$$
\begin{array}{llll}
\text{(R1)} & y_1 = x_1 - x_2 & & \le 0 \\
\text{(R2)} & y_2 = x_2 - x_3 & & \le 0 \\
\text{(R3)} & y_3 = -x_1 + x_3 & & \le 0 \\
\text{(R4)} & y_4 = x_1 + x_2 + x_3 + 10I_4 & & \le 1
\end{array}
$$

x_1, x_2, x_3 are binary {0,1}

$I_4 \ge 0$ and integer.

The ordinary bound reduction deduced that I_4=0 in every feasible solution, so the logical implications test is conducted with that already known.

From (R1), we have YMIN(1) = −1 and YMAX(1) = 1, with the extremal values XMIN = (0,1) and XMAX = (1,0), respectively. Also, YL(1) = −∞ and YU(1) = 0. For x_1 this yields:

Since $A_{11} > 0$,

```
YMIN0(1) = YMIN(1)       = -1
YMIN1(1) = YMIN(1) + 1   =  0 = YU(1)
YMAX0(1) = YMAX(1) - 1   =  0
YMAX1(1) = YMAX(1)       =  1.
```

The second rule applies: YMIN1(1) = YU(1), so we infer the logical implication:

$$\text{IF } x_1 = 1 \text{ THEN } x_2 = 1.$$

If you inspect LGLTEST.RED (after executing ATEST), you will find the following additional implications.

```
From (R2):        IF x2=1 THEN x3=1.
From (R3):        IF x3=1 THEN x2=1.
```

The contrapositives, such as `IF` $x_2 = 0$ THEN $x_1 = 0$, are purposely avoided.

The logical implications may be further processed to identify equivalence classes, forced values and logical redundancy (this is by another module, not part of ANALYZE). In this example, there is an equivalence class in that all binary variables are equal to a common variable, say z – that is, they are each 0, or they are each 1 (with the last constraint (R4), we would find they are each 0 in every feasible solution).

Chapter 10. More Views

At least in linear programming, we can solve far larger problems than we can understand.

This chapter describes more views of linear programs. The first two are graphic plots. The first of these is a bar graph plot of block numbers, which is an option with the BLOCK command. The second is a step function plot, which is an option with the SHOW command. Then, in § 10.3, the TABLE command is described, which forms and displays a (rectangular) table of values from the syntax of the linear program. In § 10.4, the GRAPH command is described and illustrated. Finally, we illustrate an view using the schema syntax.

The figures shown in the first two sections of this chapter were obtained by executing PLOTS. In general, the displays include headers (using the _TYPE command) and other commands, such as renaming blocks. You can print PLOTS.EXC to see the command specifications. Note, too, that such displays could come from a rule file.

On-line documentation of graphics capabilities is in GRAPHICS.DOC (as well as with the HELP command).

10.1 Block plots.

When displaying row or column blocks, a visual display sometimes gives greater insight than the standard tabular display. Figure 10-1 shows the output from the following sequence of commands, applied to the optimal instance of WOODNET (begin by READ P WOODNET if you want to generate this).

1. Put columns SPI with positive levels into submatrix:
   ```
   SUB COL SPI X GT L
   ```
2. Create column blocks according to location:
   ```
   BL C S SPI??
   ```
3. Display column blocks:
   ```
   BL C D
     5          COL Blocks
     Block    Mask    Number in Block
     -------------------------------
     SYN.SF   SPISF     1
     SYN.SE   SPISE     1
     SYN.LA   SPILA     1
     SYN.DE   SPIDE     2
     SYN.CH   SPICH     1
   ```

4. Rename column blocks:
```
BL C REN SYN.SF SanFran
BL C REN SYN.SE Seattle
BL C REN SYN.LA LA
BL C REN SYN.DE Denver
BL C REN SYN.CH Chicago
```

5. Type header, where //C specifies CENTER heading (relative to screen width):
```
_TYP Distribution of Pine Supply Activities with Positive Levels //C
```

6. Plot the blocks:
```
BL C P
```
  └──► Plot option

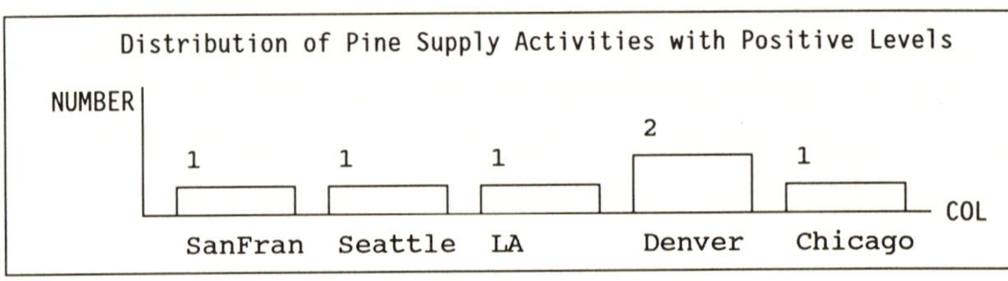

Figure 10-1. Plotting Column Blocks

The block plot is a bar graph that shows the number of activities in each column block. It contains the same information as BLOCK COLUMN DISPLAY, but it offers a visual aid that could be useful in some situations.

Let us consider another example with a linear program we have not used in other chapters. Figure 10-2 shows the schema table and syntax for a linear program, called LANDUSE (to follow along, read the packed file, LANDUSE, then read its syntax file).

```
ANALYZE ... SCH D T,D
        ::::::::::::::::::::::::: BLOCK SCHEMA :::::::::::::::::::::::::
        P(c,PR)       R(p,CR)       X(CR)          T(p,SR,DR)    RHSMODL
  B(p,PR) 1/20        -1                                         = 0
  L(PR)   1                                                      <= 1000/2200
  D(p,CR)             1             -1/1           -1/1          >= 400/5000
  COST    0.03/0.15                 0.2            0.02          ...MIN
  :LO     0           0             0              0
  :UP     0/*         *             *              *

      DOMAIN INFORMATION
   c       crop
   p       product
   PR      producing region
   CR      consuming region
   SR      consuming region
   DR      consuming region
ANALYZE ... SYN R,C

  Row syntax has 3 classes
   A row that begins with B balances some product in some producing region.
   A row that begins with L limits total land allocation in some producing
       region.
   A row that begins with D demands some product in some consuming region.

  Column syntax has 4 classes
   A column that begins with P allocates acres for some crop in some
       producing region.
   A column that begins with R receives some product in some consuming
region.
   A column that begins with X converts wheat into feed grain in some
consuming
       region.
   A column that begins with T transports some product from some consuming
       region to some consuming region.
```

Figure 10-2. An Instance of an Agricultural Planning Model (LANDUSE)

With a sequence of commands, we can plot its distribution of activity classes, as shown in figure 10-3. It shows there are 5 that convert wheat to feed grains, 100 that transport products, 20 that receive products, and 95 that allocate land in producing regions to crops. These are activities that begin with X, T, R and P, respectively.

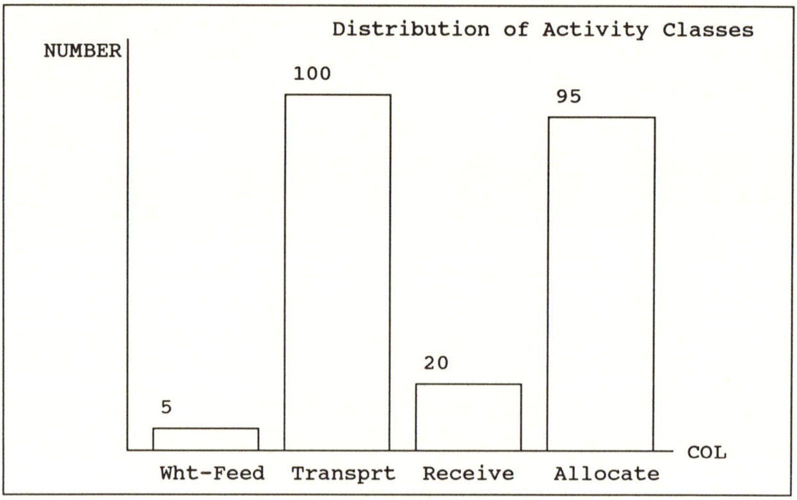

Figure 10-3. Using the BLOCK COL PLOT Command to Show Distribution of Activity Classes

Similarly, figure 10-4 shows its distribution of row classes. There are 76 balance equations, 20 demand requirements and 19 limit constraints. These are rows that begin with B, D and L, respectively.

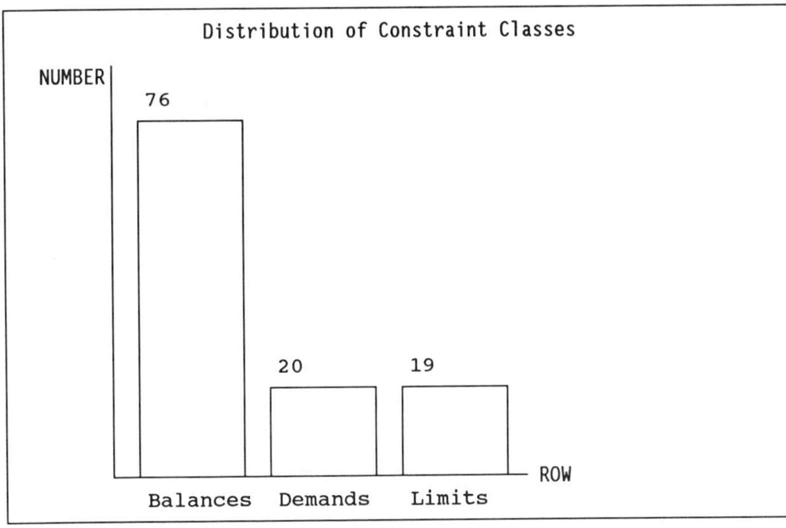

Figure 10-4. Using the BLOCK ROW PLOT Command to Show Distribution of Row Classes (excluding the objective)

You can use the SCREEN command to control the height and width of the plot. For example, figure 10-5 shows the same plot after shortening the screen dimensions by specifying SCREEN LENGTH = 15 before specifying BL R P (no header is typed this time).

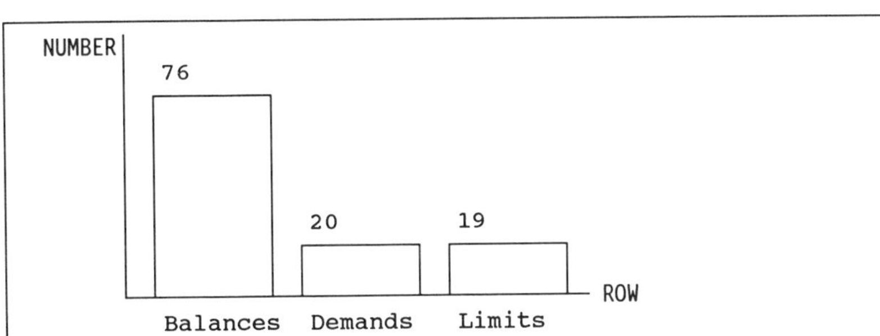

Figure 10-5. Plotting the Row Classes with Screen Length = 15

Further, you can specify that the characters be printable. Figure 10-6 illustrates this. This option is useful when the output is spooled to the printer or to a file for printing. The graphic characters might be printable, depending on your printer and on your word processor, in which case you should not use this option. The //PRINTABL specification results in the use of standard ASCII characters, which are always printable, even with the DOS PRINT command.

```
ANALYZE ... BL R P //PRINTABL
  NUMBER|
        |  76
        | .-------.
        | |       |
        | |       |
        | |       |
        | |       |  20      19
        | |       | .-------. .-------.
        | |       | |       | |       |
        \-^-------^-^-------^-^-------^- ROW
           Balances  Demands   Limits
```

Figure 10-6. Using Printable Characters for the Plot of Block Numbers

Figure 10-7 shows the plot of activity classes for those with positive levels. Note there are no activities that convert wheat to feed grains in this solution (i.e., the levels of activities whose name begins with X are all zero).

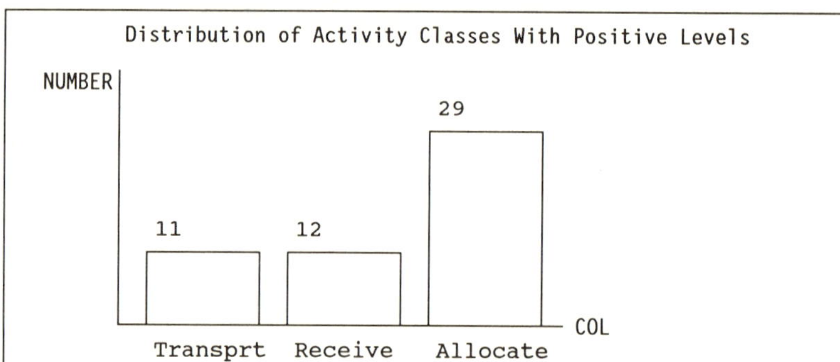

Figure 10-7. A Plot of Activity Class Distribution Over Those That Have Positive Levels in This Instance

10.2 Step plots.

Suppose we have a collection of bounded activities with positive coefficients in a particular row, such as the objective. In particular, suppose $0 \le x_j \le U_j$ and $A_{ij} > 0$ for $j=1,...,n$ (where i could be an index for the objective row). Then, we define the following step function for $0 \le z \le \Sigma_j U_j$.

$$
\begin{array}{lll}
f(0) = 0 & & \\
f(z) = A_{i1} & \text{for } 0 & < z \le U_1 \\
f(z) = A_{i2} & \text{for } U_1 & < z \le U_2 \\
f(z) = A_{i3} & \text{for } U_2 & < z \le U_3 \\
\vdots & & \\
f(z) = A_{in} & \text{for } U_{n-1} & < z \le U_n
\end{array}
$$

The SHOW command has a PLOT option that plots this step function for submatrix columns over each submatrix row. Figure 10-8 displays activities in the optimal instance of WOODNET that supply pine (i.e., those whose name begins with SPI and has a positive level in the solution). Then, their cost values are listed.

```
ANALYZE ... D C SPI X GT L //LO,UP
  6 Columns are in submatrix.
  COL          LO_BOUND      UP_BOUND
  ---------------------------------
  SPICH1             0        100.000
  SPIDE1             0        100.000
  SPIDE2             0        200.000
  SPILA1             0         50.000
  SPISE1             0         10.000
  SPISF1             0         90.000
ANALYZE ... L COST
       MIN     COST     =    11 SPICH1 +   10 SPIDE1 +   12 SPIDE2
                        +    15 SPILA1 +   11 SPISE1 +   12 SPISF1
```

Figure 10-8. Pine Supply Activities in WOODNET with Positive Levels

Figure 10-9 illustrates SHOW PLOT specification, which plots the step function associated with these six supply activities over the objective row (COST). The NOSORT specification plots the step for the columns in their natural order, as from the DISPLAY command. Activity SPICH1 has COST coefficient = 11, so this is the height of the first step. Its bound value is 100, so this is the width of the first step. Step 2 is associated with activity SPIDE1, whose COST coefficient = 10 and whose bound value is also 100. The fifth step has a very small width. It is associated with activity SPISE1, which has a bound value of only 10.

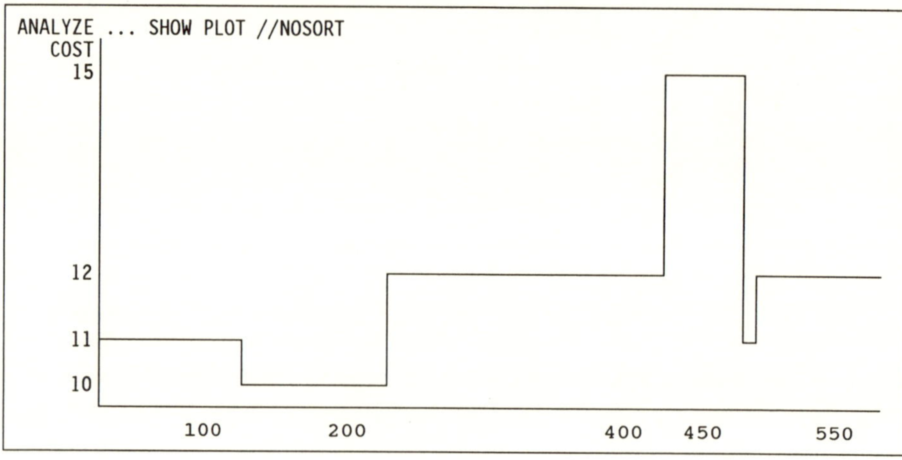

Figure 10-9. Plotting the Step Function of Pine Supply Activities in WOODNET with Positive Levels Over the Objective Row (COST)

With step functions, especially over the objective row, it is desirable to sort the steps in ascending order. Figure 10-10 shows this is the default (i.e., not specifying NOSORT as an option). The small blips you see are the steps with the same height that merge into each other. In particular, activities SPICH1 and SPISE1 both have COST coefficient = 11; and, activities SPIDE2 and SPISF1 both have COST coefficient = 12 (c.f., figure 10-8).

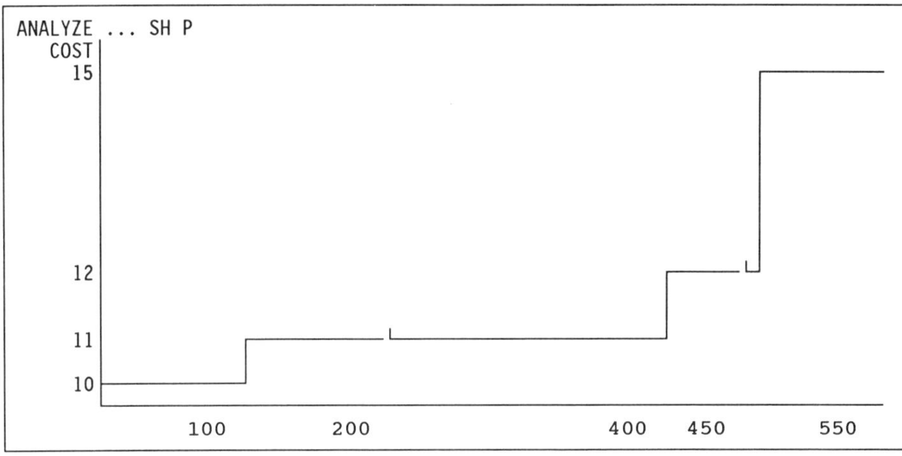

Figure 10-10. Showing the Step Function in Sort Order

As with the block plot, the screen width and height can be changed to change the appearance of the step function. In addition, printable characters can be specified in the same manner as before.

10.3 Table formations and displays.

The TABLE command gives another view of 2-dimensional information. To illustrate, consider the display in figure 10-11. This was done after reading WOODNET packed and syntax files, then displaying its schema. To follow, enter the following commands.

```
READ P WOODNET
READ S
SCH  D
D C T**SF
```

You should now see the same display as in figure 10-11.

This is a standard way to see rim and solution values for the activities that represent transportation of materials out of San Francisco.

```
ANALYZE ...D C T**SF
  16 Columns are in submatrix.
  COL       STAT  LEVEL (X)     LO_BOUND      UP_BOUND      PRICE (D)
  ------------------------------------------------------------------
  TMOSFCH    L        0             0             *            5.000
  TMOSFDE    B        0             0             *            0
  TMOSFLA    B       20.000         0             *            0
  TMOSFSE    B       10.000         0             *            0
  TOKSFCH    L        0             0             *           12.000
  TOKSFDE    B       10.000         0             *            0
  TOKSFLA    B        5.000         0             *            0
  TOKSFSE    B        5.000         0             *            0
  TPISFCH    L        0             0             *            5.000
  TPISFDE    L        0             0             *            2.000
  TPISFLA    B       50.000         0             *            0
  TPISFSE    B       10.000         0             *            0
  TWASFCH    L        0             0             *            7.000
  TWASFDE    L        0             0             *           14.000
  TWASFLA    B        5.000         0             *            0
  TWASFSE    B        5.000         0             *            0
```

Figure 10-11. Displaying Transportation Activities To See Levels and Prices Out of San Francisco

Figure 10-12 illustrates the TABLE command to see a rectangular display of levels and prices. The command syntax is:

TABLE {ROW | COL} *name key*

where *name* is the class name and *key* is the usual designation of level (X for column and Y for row), objective coefficient (C, for column only), price (D for column and P for row), lower bound (L), or upper bound (U). In the first table specification the class name is T and the key is X. In the second, the key is D.

The first line of the table display describes what is displayed, using the schema syntax to show the class and domain. Since the set of supply regions (SR) is fixed at member = SF (San Francisco), it is shown as part of this table header. Then, set DR is identified as the table's column heads.

The next line gives the table's stub set, which is MT (set of materials), followed by the column head members that vary in this domain. That is, set DR has members CH, DE, LA and SE, shown as the table's column heads, which are in the domain of activity T.

Each subsequent line begins with the member of the stub set, starting with MO (mahogany). Then, the values appear as entries in the table.

```
ANALYZE ...TAB C T X
                            Levels of T(MT,SR,DR) for SR=SF
                            =============== DR ===============
                        MT  CH        DE        LA        SE
                         MO 0         0         20        10
                         OK 0         10        5         5
                         PI 0         0         50        10
                         WA 0         0         5         5
ANALYZE ...TAB C T D
                            Prices of T(MT,SR,DR) for SR=SF
                            =============== DR ===============
                        MT  CH        DE        LA        SE
                         MO 5         0         0         0
                         OK 12        0         0         0
                         PI 5         2         0         0
                         WA 7         14        0         0
```

Figure 10-12. Using the TABLE Command to See Levels and Prices of Transportation out of San Francisco in Rectangular Format

Now suppose we are interested in the distribution of pine. These are the levels of the transportation activities, TPI. Figure 10-13 shows this for all such activities, including those with zero level.

```
ANALYZE ...SUB COL TPI
ANALYZE ...TAB C T X
                            Levels of T(MT,SR,DR) for MT=PI
                       ==================== DR ===================
                   SR  CH        DE        LA        SE        SF
                    CH           100       0                   0
                    DE 100                 0         0         30
                    LA 0         0                             50
                    SE 0         0         0                   10
                    SF 0         0         50        10
```

Figure 10-13. Displaying the Distribution of Pine

The blanks mean the associated activity does not exist in the submatrix. Since we set the submatrix to contain all such columns, they do not exist in the linear program. In particular, there are no transportation links from CH to CH, from CH to SE, from DE to DE, from LA to LA, from LA to SE, from SE to SE, and from SF to SF.

Figure 10-14 shows an alternative table formation. First, the activities with zero levels are deleted from the submatrix. Then, the blanks in the table display still mean those activities are absent from the submatrix. The links could still be in the linear program; they simply have zero level in this solution.

```
ANALYZE ...SUB D C * X=0
ANALYZE ...TAB C T X
                        Levels of T(MT,SR,DR) for MT=PI
                  ==================== SR ===================
               DR  CH        DE         LA         SE         SF
                CH           100
                DE 100
                LA                                            50
                SE                                            10
                SF           30         50         10
```

Figure 10-14. Displaying only Positive Transportation Levels of Pine

Another thing happened - the table stubs are demand regions (DR), and the table heads are supply regions (SR). This happened because the first set to change value in the domain of the activities is the set SR. Use the DISPLAY command to verify this. Shortly, we shall describe defaults for stub and head formation and how to over-ride them.

Figure 10-15 displays all supply columns. This column class has exactly two domain sets (material and region), both of which vary in the submatrix.

```
ANALYZE ...D C S
  18 Columns are in submatrix.
  COL       STAT  LEVEL (X)    LO_BOUND     UP_BOUND      PRICE (D)
  -------------------------------------------------------------------
  SMOSE1     B       75.000        0            *              0
  SMOSF1     U       30.000        0          30.000     -  18.000
  SOKCH1     U       40.000        0          40.000     -  23.000
  SOKDE1     U       50.000        0          50.000     -  20.000
  SOKLA1     U        5.000        0           5.000     -  15.000
  SOKSE1     B       15.000        0            *              0
  SOKSF1     U       20.000        0          20.000     -  17.000
  SPICH1     B      100.000        0         100.000         0
  SPICH2     L        0            0         100.000          4.000
  SPIDE1     U      100.000        0         100.000     -    2.000
  SPIDE2     B       30.000        0         200.000         0
  SPILA1     U       50.000        0          50.000     -    4.000
  SPISE1     U       10.000        0          10.000     -    3.000
  SPISF1     B       60.000        0          90.000         0
  SPISF2     L        0            0            *            0
  SWACH1     U       25.000        0          25.000     -    7.000
  SWASE1     B       40.000        0            *            0
  SWASF1     U       10.000        0          10.000     -  20.000
```

Figure 10-15. Displaying all Supply Activities

See what happens, however, when a table is displayed in figure 10-16. The table of levels is formed, but there are multiple entries, as indicated by the footnote.

```
ANALYZE ...TABL C S X
                                 Levels of S(MT,SR)
                      ==================== SR ===================
                 MT   CH         DE         LA         SE        SF
                  MO                                   75        30
                  OK  40         50         5          15        20
                  PI  100        130#       50         10        60
                  WA  25                               40        10
                        # multiple entries, which were added.
```

Figure 10-16. Table of Supply Activity Levels

This occurs because there is one more character in each name (1 or 2 at the end), which is not specified in the syntax. The domain of S is MT×SR, which is

2-dimensional, but activities SPIDE1 and SPIDE2 create multiple entries for the table entry at MT=PI and SR=DE. Examine figure 10-15 and note: the level of SPIDE1 is 100, and the level of SPIDE2 is 30. These are added to yield the table entry, 130.

The other multiple entries, such as SPICH1 and SPICH2, do not create an ambiguity because one of the levels is zero. The footnote does not appear for such cases. The footnote mark (#) appears for table entries of quantities only when adding nonzero quantities to form the table entry.

Figure 10-17 shows the table of upper bounds, where the footnote applies to S(PI,SF), S(PI,CH), and S(PI,DE). Examine figure 10-15 to see the upper bounds for SPISF1, SPISF2, etc.

```
ANALYZE ...TAB C S U
                         Upper bounds of S(MT,SR)
              ==================== SR ===================
          MT  CH        DE        LA        SE        SF
          MO                                *         30
          OK  40        50        5         *         20
          PI  200#      300#      50        10        *#
          WA  25                            *         10
                # multiple entries, which were added.
```

Figure 10-17. Table of Supply Upper Bounds

Figure 10-18 shows the same table with the stub and head switched. By using the FOR option, the first set specified (SR) is the stub and the second (MT) is the head. The same table would appear if we specified only FOR SR (omitting the head specification, MT), because MT is the only other set in the domain, so it must be the head once SR is defined to be the stub.

```
ANALYZE ...TAB C S U FOR SR,MT
                         Upper bounds of S(MT,SR)
                     =============== MT ===============
                SR  MO        OK        PI        WA
                 CH           40        200#      25
                 DE           50        300#
                 LA           5         50
                 SE  *        *         10        *
                 SF  30       20        *#        10
                  # multiple entries, which were added.
```

Figure 10-18. Reversing Stub and Head Using the FOR option

Figure 10-19 shows what happens when we display the table of cost coefficients. Now the footnote is a bit different. When there are multiple entries of prices, the action taken is not to add them, but to enter the one with maximum value. If the values are the same, no footnote mark appears; otherwise, the footnote is given as shown. Further, the footnote mark for costs is ?, rather than #.

The COST values are listed for you to compare with the table entries. The multiple entries, SPICH1 and SPICH2 have costs 11 and 15, respectively. The 15 is entered in the table with the footnote mark. On the other hand, SPISF1 and SPISF2 have the same cost (12), so no footnote mark appears.

```
ANALYZE ...TAB C S C
                      Objective coefficients of S(MT,SR)
                 =================== SR ===================
             MT  CH        DE        LA        SE        SF
              MO                                55        45
              OK  35        30        40        50        40
              PI  15?       12?       15        11        12
              WA  50                            60        50
                   ? multiple entries; the max is shown.
ANALYZE ...L COST
    MIN   COST   =    55 SMOSE1 +  45 SMOSF1 +  35 SOKCH1 +  30 SOKDE1
                 +    40 SOKLA1 +  50 SOKSE1 +  40 SOKSF1 +  11 SPICH1
                 +    15 SPICH2 +  10 SPIDE1 +  12 SPIDE2 +  15 SPILA1
                 +    11 SPISE1 +  12 SPISF1 +  12 SPISF2 +  50 SWACH1
                 +    60 SWASE1 +  50 SWASF1
```

Figure 10-19. Table of Objective Coefficients (Costs) for Supply Activities

Figure 10-20 shows what happens when we display the table of prices (or reduced costs). Now the action is similar to costs, except the entry with maximum absolute value is entered. As before, no footnote mark is placed if the multiple entries have the same prices. For example, SPISF1 and SPISF2 both have zero price, so there is no footnote mark with the entry. The entry for S(PI,DE) is –2, and it has a footnote to indicate there is another activity in this class with MT=PI and SR=DE that has a price whose magnitude is less than 2. Examine figure 10-15 and note that the price of SPIDE2 is 0.

```
ANALYZE ...TAB C S D
                          Prices of S(MT,SR)
          ==================== SR ===================
      MT  CH        DE        LA        SE        SF
      MO                                0         -18
      OK  -23       -20       -15       0         -17
      PI  4?        -2?       -4        -3        0
      WA  -7                            0         -20
       ? multiple entries; the max magnitude is shown.
```

Figure 10-20. Table of Prices (i.e., Reduced Costs) of Supply Activities

Multiple entries are also possible when the domain of the class has more than two sets that vary. For example, figure 10-21 shows the table formed for all transportation activities, which has domain MT×SR×DR. The stub and head are left to default sets, which are determined as follows. The first set to vary in the submatrix is the head, and the second is the stub. Due to the sort order of the activities (by name), set DR changes first and set SR changes second. The table title specifies MT=*, which means set MT also varies in the submatrix, so the table entries are total levels wherever the footnote appears. For example, the total levels of transportation from DE to CH is 140, and the footnote means more than one material has a positive level. On the other hand, transportation from CH to LA is 5 with no footnote mark. The absence of the footnote means only one material has positive level, which is 5. Similarly, the levels of transportation from CH to SF are all zero, indicated by the 0 entry with no footnote.

```
ANALYZE ...SUB *
ANALYZE ...TAB C T X
                              Levels of T(MT,SR,DR) for MT=*
                         =================== DR ===================
                     SR  CH         DE         LA         SE         SF
                     CH             160#       5                     0
                     DE  140#                  0          10         30
                     LA  0          0                                55#
                     SE  50#        20         5                     65#
                     SF  0          10         80#        30#
                           # multiple entries, which were added.
```

Figure 10-21. Table of Aggregate Transportation Levels, Summed over Materials

By using the FOR option, you can specify which two of the three sets will be the stub and head. Figure 10-22 illustrates this by specifying the set of supply regions (SR) as the stub and the set of materials (MT) as the head. Now the entries with footnotes are aggregate levels, summed over demand regions (DR).

```
ANALYZE ...TAB C T X FOR SR,MT
                              Levels of T(MT,SR,DR) for DR=*
                             =============== MT ===============
                         SR  MO         OK         PI         WA
                         CH             40         100        25#
                         DE             50#        130#
                         LA             5          50
                         SE  75#        15         10         40#
                         SF  30#        20#        60#        10#
                           # multiple entries, which were added.
```

Figure 10-22. Table of Aggregate Transportation Levels, Summed over Demand Regions

The examples so far illustrate the TABLE command with one row or column class over two sets that vary in the submatrix. Now consider the formation and display of a table whose heads are row or column classes and whose stubs are members of one varying set that is in the domain of each class.

Figure 10-23 shows the specification of multiple classes, in this case supply and demand rows, whose names begin with S and D, respectively. They have exactly one set, that is in both domains, namely MT (materials). This is called their common set and defines the stubs of the table. The row classes, S and D, comprise the table heads. The table entries are prices, specified by the key P.

```
TABLE ROW  S,D P
           └┬┘ └──▸ key = P means entries = prices
            └─────▸ row classes, separated by commas
```

```
ANALYZE ...TABLE ROW  S,D P
                      Table Entries = (Absolute Max) Prices
                             MT   S         D
                              MO  63        75
                              OK  58        69
                              PI  19        24
                              WA  70        78
```

Figure 10-23. Displaying a Table of Supply and Demand Row Prices

The reference, '(Absolute Max)', in the table's title, means the same as in the case of a single class: the values displayed are those whose magnitudes are greatest if there are multiple entries. For example, rows SMOSE, SMOSF, SMOCH, SMODE and SMOLA are members of S(MT,SR) for MT=MO. If you display these rows, you will see that the price with greatest magnitude is 63, as shown.

Unlike the display of a single class, no footnote is given in the display of multiple classes. The rules, however, for multiple entries are the same. Quantities (levels and bounds) are added, and prices (and objective coefficients) are maximized.

Figure 10-24 shows a similar table for the solution prices of the three column classes, where their common set is also MT. The activity supply prices are negative, and the most negative is the table entry.

```
ANALYZE ...TAB C S,T,D D
                      Table Entries = (Absolute Max) Prices
                        MT   S          T          D
                         MO  -18        5          75
                         OK  -23        14         69
                         PI  -4         13         24
                         WA  -20        14         78
```

Figure 10-24. Table Display of Prices (Reduced Costs) of All Column Classes

Now consider the attempt to display classes S and T. Their domains are MT×SR and MT×SR×DR, respectively, so they have two sets in common, MT and SR. The error message given in figure 10-25 is due to this ambiguity. Then, this is resolved by the use of the FOR option, which specifies which set to use as the common set. The first table display is for the set of materials (MT), and the second is for the set of supply regions (SR).

```
ANALYZE ...TAB C S,T C
ANALYZE ...** COMMON SET NOT DEFINED...USE FOR
           ...TABLE ABORTED
ANALYZE ...TAB C S,T C FOR MT
                  Table Entries = (Max) Objective coefficients
                               MT   S        T
                                MO  55       20
                                OK  50       20
                                PI  15       20
                                WA  60       20
ANALYZE ...TAB C S,T C FOR SR
                  Table Entries = (Max) Objective coefficients
                               SR   S        T
                                SE  60       20
                                SF  50       15
                                CH  50       18
                                DE  30       15
                                LA  40       16
```

Figure 10-25. Using the FOR Option to Specify a Common Set for Multiple Class

Try similar cases for rows and columns of other linear programs. You might also look at REPORT.RUL, which is a rule file that uses the TABLE command to produce a report for three linear programs: WOODNET, BLENDRT and FOREST5. To instantiate this rule file, specify INTERPRT TEST after setting STRING TEST = REPORT. Then, look at the output file(s) created: WOODNET.RPT, BLENDRT.RPT and FOREST.RPT, respectively.

10.4 The GRAPH command.

The GRAPH command offers another view of a submatrix. To illustrate, figure 10-26 pictures the submatrix associated with positive distribution of mahogany (MO).

```
ANALYZE ...  PIC *MO,* X=1/*
                             Picture of submatrix
          D D D D D S S T T T T T
          M M M M M M M M M M M M
          O O O O O O O O O O O O
          C D L S S S S S S S S S
          H E A E F E F E E E F F
          1 1 1 1 1 1 1 C D S L S
                        H E F A E
 DMOCH    -             +         =  0
 DMODE      -             +       =  0
 DMOLA        -               +   =  0
 DMOSE          -               + =  0
 DMOSF            -         +     =  0
 SMOSE              +   - - -     =  0
 SMOSF                +       - - =  0
```

Figure 10-26. Setting the Graphs for Mahogany Distribution in WOODNET

The sign pattern of the picture induces a directed bigraph, called the *fundamental digraph*, defined as follows. There is a node for every row and for every column. Each nonzero in the submatrix corresponds to a link between the row and column nodes, oriented as:

$$(\text{row } i) \xrightarrow[A_{ij} < 0]{} [\text{column } j] \xrightarrow[A_{kj} > 0]{} (\text{row } k)$$

It is useful to think of a positive coefficient as an output of the activity, and a negative coefficient as an input to the activity. In WOODNET, each supply activity (whose name begins with S) has a single positive coefficient, which is in the associated supply balance row; and, each demand activity (whose name begins with D) has a single negative coefficient, which is in the associated demand balance row. Each transportation activity (whose name begins with T) has the associated supply row as an input (i.e., negative coefficient) and the associated demand row as an output (i.e., positive coefficient).

The fundamental digraph has two projections that give insight into relational structures in the linear program. The first is the *row digraph*. This has a node for every row and arcs as follows. There is an arc $<i, k>$, from row i to row k, if there is a column with row i as output and row k as input – that is, there exists j such that

$A_{ij} < 0$ and $A_{kj} > 0$. An example is the transportation activity that links a supply row to a demand row. In addition, a headless arc, of the form (i)——▸, emanates from the node associated with row i if there is a non-positive column, say j, such that $A_{ij} < 0$. The demand activities, for example, induce headless arcs on their associated demand rows, which indicates a flow requirement. A tail-less arc, of the form ——▸(i), is an input to row node i if there is a non-negative column, say j, such that $A_{ij} > 0$. The supply activities, for example, induce tail-less arcs on their associated supply rows. Non-zero right-hand sides, which represent supply inputs or demand outputs also induce tail-less and headless arcs, respectively, as part of the row digraph's construction.

The second projection is the *column digraph*. This construction is defined similar to the row digraph, where there is a node for every column, and an arc from one column to another if there is a row that connects them with a 2-step (directed) path.

Figure 10-27 specifies GRAPH SET * to set the bigraph and both the row digraph and the column digraph. It then shows the display of the row digraph. The format of the display, called a *diagram*, has the form:

predecessor ——▸ row ——▸ successor

This is done for each row node. Multiple predecessors and successors appear on separate lines.

```
ANALYZE ...  GRAPH SET *
ANALYZE ...  GRAPH DISP ROW
```

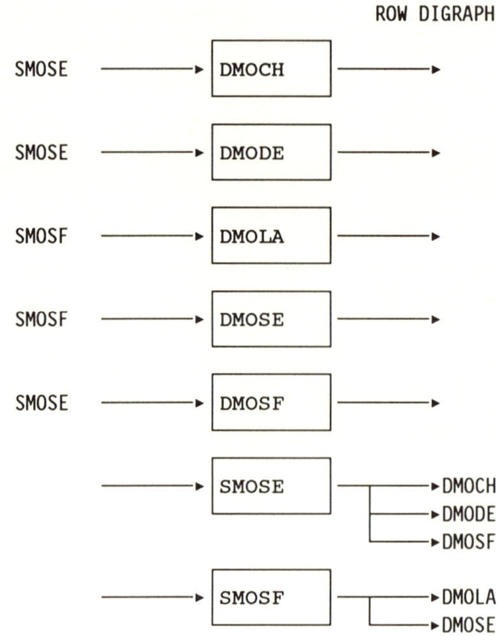

Figure 10-27. Row Digraph for Mahogany Distribution in WOODNET

The first row node shown is DMOCH, which has a headless arc to represent the demand requirement associated with column DMOCH1 in the submatrix (c.f., figure 10-26). Its predecessor is row node SMOSE, which is induced by the associated transportation activity, namely column TMOSECH. This structure is the same for each of the five demand row nodes shown.

The row node SMOSE has a tail-less arc to represent supply input associated with column SMOSE1 in the submatrix. It has three successors, namely rows DMOCH, DMODE, and DMOSF. These arcs are induced by the associated transportation activities, and the display shows where each supplier sends mahogany.

The arrows display smoothly on your screen, but they will not print the same way (unless you load the right font). The DIAGRAM option allows another option: GRAPHIC or PRINTABL. The former is the default, which is what is shown in figure 10-27; the latter is shown in figure 10-28. The PRINTABL option is useful when spooling output to a file for documentation; the GRAPHIC output may not be printable.

```
ANALYZE ... GR D R //DIAGRAM PRINTABL
                                ROW DIGRAPH
                   .--------.
 SMOSE    ------->|DMOCH    |------->
                   \--------/
                   .--------.
 SMOSE    ------->|DMODE    |------->
                   \--------/
                   .--------.
 SMOSF    ------->|DMOLA    |------->
                   \--------/
                   .--------.
 SMOSF    ------->|DMOSE    |------->
                   \--------/
                   .--------.
 SMOSE    ------->|DMOSF    |------->
                   \--------/
                   .--------.
          ------->|SMOSE    |--.---->DMOCH
                   \--------/  :---->DMODE
                               \---->DMOSF
                   .--------.
          ------->|SMOSF    |--.---->DMOLA
                   \--------/  \---->DMOSE
```

Figure 10-28. Specifying Printable Characters

Figure 10-29 shows a matrix format option for the graph display. Here the adjacency matrix is shown, where a 1 indicates an arc. The diagonal entries, which could be +, –, or * represent headless (output) and tail-less (input) arcs, respectively (if both arcs are incident with a node, the adjacency matrix contains * in its diagonal).

```
ANALYZE ...  GRAPH D R //MATRIX
                                            ROW DIGRAPH
             01 02 03 04 05 06 07
  01 DMOCH   +
  02 DMODE      +
  03 DMOLA         +
  04 DMOSE            +
  05 DMOSF               +
  06 SMOSE   1  1        1  -
  07 SMOSF         1  1        -
```

Figure 10-29. The Adjacency Matrix Display Option

The GRAPH command does not perform any operations on the graph or its projections. It is, therefore, sometimes useful to separate it into components. This can be done with the BLOCK command, as shown in figure 10-30. After blocking the submatrix and finding two components, the submatrix is cleared with the COUNT command (C .). Then, the first component, called COMP.001 for both the row and column blocks, is copied over to comprise the submatrix. The picture shows we have one supplier and three demands.

```
ANALYZE ...  BLOCK

          2 components found
ANALYZE ...  C .
NO ROW MATCHES .
   Submatrix has  0 rows, 0 columns, and 0 nonzeroes
ANALYZE ...  BL R C F COMP.001
ANALYZE ...  BL C C F COMP.001
ANALYZE ...  PIC
                                  Picture of submatrix
          D D D S T T T
          M M M M M M M
          0 0 0 0 0 0 0
          C D S S S S S
          H E F E E E E
          1 1 1 1 C D S
                  H E F
 DMOCH    -       +     =  0
 DMODE      -       +   =  0
 DMOSF        -       + =  0
 SMOSE          + - - - =  0
```

Figure 10-30. Blocking the Graph into Components

In figure 10-31, the new graph is set, and the row digraph is displayed. You might repeat this for the second component and display the column digraph. Try the MATRIX option to see the adjacency matrix in each case.

```
ANALYZE ...  GR SET *
ANALYZE ...  GR D R
```

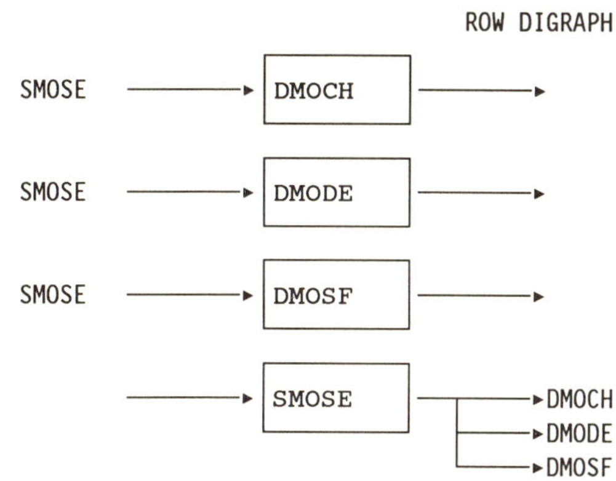

Figure 10-31. The Row Digraph of a Component

Figure 10-32 shows the result of the syntax option. Because mahogany (MO) is the only member of the set of materials (MT) in the domains of rows, it appears in the header. The same applies to Seattle (SE) being the only supply region (member of SR). Then, each remaining set in the row syntax is translated, giving flows from supply regions to demand regions. In this case, the tail of each arc is blank because the link is from the only supply row, SMOSE, and all of its domain entries are in the header.

The header can be over-ridden by specifying NOHEADER after the SYNTAX option. As shown in figure 10-32, this translates all domains of each row node.

```
ANALYZE ... GR D R //SYNTAX
                                           ROW DIGRAPH
                                 Flow for MT=mahogany , SR=Seattle
   --->Chicago
   --->Denver
   --->San Francisco
ANALYZE ...GR D R //SYN NOHEADER
                                           ROW DIGRAPH
  mahogany Seattle --->mahogany Chicago
                   --->mahogany Denver
                   --->mahogany San Francisco
```

Figure 10-32. Syntax Format of Row Digraph

The headless and tail-less arcs are not displayed with the syntax option. The arcs shown correspond to activities in the submatrix that transport mahogany from one place to another.

Consider another exercise with an instance of a blending model. Figure 10-33 displays the schema after reading in the packed file (matrix + solution previously read and written) and the associated syntax file (BLENDRT.SYN). Note there are sets that are *Attributed*, which we shall explain shortly.

```
(after READ P BLENDRT and READ SYN)
ANALYZE ...SCHEMA DISP TABLE,DOM
           :::::::::::::::::::::: BLOCK SCHEMA ::::::::::::::::::::::
           P(MT,R,T)   O(UN,R,T)   T(MT,S,D,T) I(MT,R,T)   RHSMODL
 B(MT,R,T) 1           -1/118      -1/1        -1/1        >= 0/260
 COST      30/36       5/13        10          0.01/0.1    ...MIN
 :LO       0           0           0           0/10
 :UP       *           100/300     *           *

     DOMAIN INFORMATION
  UN       process unit
  MT       material (Attributed FORM)
  R        process region (Attributed PLACE)
  S        source region (Attributed PLACE)
  D        destination region (Attributed PLACE)
  T        time period (Attributed TIME)
```

Figure 10-33. A Blending Model Instance (BLENDRT)

Figure 10-34 pictures the submatrix composed of balance rows for unleaded regular, whose names all begin with BUR, and activities whose levels are greater than their lower bounds. To gain insight into unleaded regular, we first look at the row digraph of this submatrix, as shown.

```
ANALYZE ...PI BUR,* X GT L
                               Picture of submatrix
         I I O O T T T
         U U 1 1 U U U
         R R 1 1 R R R
         1 1 1 1 1 1 1
         1 2 1 2 2 2 2
                 1 2 3
 BUR11   -   +   -     > +
 BUR12   + -   +   -   > +
 BUR13     +         - > +
 BUR21           +     > +
 BUR22             +   > +
 BUR23               + > +
ANALYZE ...GRAPH SET *
ANALYZE ...GRAPH DISPLAY ROW //SYNTAX
                                     ROW DIGRAPH
                              Flow for MT=unleaded regular
  Texas 1st time period --->Texas 2nd time period
                        --->Louisiana 1st time period
  Texas 2nd time period --->Texas 3rd time period
                        --->Louisiana 2nd time period
  Texas 3rd time period --->Louisiana 3rd time period
```

Figure 10-34. A Picture of the Balance Rows for Unleaded Regular with Activities Whose Levels are Above Their Lower Bounds (zero)

Now we trace a margin-setting path, as shown in figure 10-35. The picture shows an enlarged submatrix to balance the flow of materials and products over space and time.

```
ANALYZE ...TRACE
 TRACE complete...submatrix is set
ANALYZE ...PI
                                          Picture of submatrix
          I I O O O O P P T T T
          U U 0 0 1 1 H H U U U
          R R 1 1 1 1 . . R R R
          1 1 1 1 1 1 1 1 1 1 1
          1 2 1 2 1 2 1 2 2 2 2
                          1 2 3
 BH.11        -       +         > 0
 BH.12          -       +       > 0
 BRF11        +   -             > 0
 BRF12          +   -           > 0
 BUM11        +                 > +
 BUM12          +               > +
 BUR11    -       +       -     > +
 BUR12    + -       +       -   > +
 BUR13      +                 - > +
 BUR21                    +     > +
 BUR22                      +   > +
 BUR23                        + > +
 COST     + + + + + + + + + + + = MIN
```

Figure 10-35. Tracing a Margin-Setting Path for Unleaded Regular

Figure 10-36 shows the row digraph for this submatrix, using the syntax to translate each row into English. The first entry is the arc from row BH.11, which translates to 'heavy crude in Texas at 1st time period', to row BRF11, which translates to 'raffinate in Texas at 1st time period'. This arc is due to the operation activity, O0111, which operates a primary unit in Texas at the first time period. This operation activity uses heavy crude as an input and produces raffinate as one of its outputs. The second arc comes from the fact that this same activity also produces unleaded mid grade as another output.

In general, each operation activity (which begins with O) is a *transformation of form* – that is, it has some material as an input and produces another material as an output (a primary unit uses crude oil as an input and produces some final product as well as blend stocks; a blend unit uses blend stocks as inputs and produces some final product).

```
ANALYZE ...GRAPH SET *
ANALYZE ...GRAPH DISP ROW //SYN
                                   ROW DIGRAPH
 heavy crude Texas 1st time period --->raffinate Texas 1st time period
          --->unleaded mid grade Texas 1st time period
 heavy crude Texas 2nd time period --->raffinate Texas 2nd time period
          --->unleaded mid grade Texas 2nd time period
 raffinate Texas 1st time period --->unleaded regular Texas 1st time period
 raffinate Texas 2nd time period --->unleaded regular Texas 2nd time period
 unleaded regular Texas 1st time period
          --->unleaded regular Texas 2nd time period
          --->unleaded regular Louisiana 1st time period
 unleaded regular Texas 2nd time period
          --->unleaded regular Texas 3rd time period
          --->unleaded regular Louisiana 2nd time period
 unleaded regular Texas 3rd time period
          --->unleaded regular Louisiana 3rd time period
```

Figure 10-36. Row Digraph of Submatrix in Figure 10-35 with English Translations of Rows

An inventory activity (which begins with I) is a *transformation of time* – that is, it has some time period as an input and produces the same material at the same location in the next time period as an output. The seventh arc,

```
unleaded regular Texas 1st time period
          --->unleaded regular Texas 2nd time period
```

is due to inventory activity, IUR11. This links row BUR11, which translates to 'unleaded regular in Texas at 1st time period', to row BUR12, which translates to 'unleaded regular in Texas at 2nd time period'. The output shown excludes the prepositions ('in' and 'at'), but otherwise gives the transformation as a link in the row digraph, using English phrases from the syntax.

A transportation activity (which begins with T) is a *transformation of place* – that is, it has some region as an input and produces the same material at the same time period at another region as an output. The activity TUR121 transforms the location of unleaded regular in Texas at 1st time period to Louisiana. Graphically, this links row BUR11 to row BUR21, as shown in figure 10-36.

These transformations of form, place and time (introduced by F.H. Murphy and E.A. Stohr) can be delineated by using the set *attribution*. The syntax file, BLENDRT.SYN, specifies the set attributions (which was created by MODLER in this case). Before we demonstrate the delineations by attribution, save this graph, as shown in figure 10-37.

After saving the graph, figure 10-37 displays the row digraph when activities are restricted to be transformations of form. Notice the headers that are formed for sets that do not change – that is, the activities do not transform their values in the subgraph delineated by the attributes. If you picture the submatrix associated with this digraph (by entering PIC at the ANALYZE prompt), you will see that it has precisely the four operation activities (whose names begin with O), which are the only transformations of form in the original submatrix. The 7 deleted columns are the 2 Inventory, 2 Production and 3 Transportation activities.

```
ANALYZE ... GR SAVE
ANALYZE ... GR SET * //FORM
                   7 Columns deleted from submatrix
ANALYZE ... GR D R //S
                                          ROW DIGRAPH
                                        Flow for R=Texas
 heavy crude 1st time period --->raffinate 1st time period
                             --->unleaded mid grade 1st time period
 heavy crude 2nd time period --->raffinate 2nd time period
                             --->unleaded mid grade 2nd time period
 raffinate 1st time period --->unleaded regular 1st time period
 raffinate 2nd time period --->unleaded regular 2nd time period
```

Figure 10-37. Setting and Displaying Row Digraph for Transformations of Form

Now load the original graph, then copy the graph to the submatrix. These are the first two commands shown in figure 10-38. Next, the graph is set with activities that are transformation of place. If you now picture the submatrix, you will find precisely three transportation activities, which are the only transformations of place in the original submatrix.

```
ANALYZE ... GR LOAD
ANALYZE ... GR COPY
ANALYZE ... GR SET * //PLACE
                        8 Columns deleted from submatrix
ANALYZE ... GR D R //S
                                          ROW DIGRAPH
                                  Flow for MT=unleaded regular
  Texas 1st time period --->Louisiana 1st time period
  Texas 2nd time period --->Louisiana 2nd time period
  Texas 3rd time period --->Louisiana 3rd time period
```

Figure 10-38. Setting and Displaying Row Digraph for Transformations of Place

Finally, load the graph again and copy it to the submatrix, as shown in figure 10-39. Then, set the graph for activities that are transformations of time. These are precisely the inventory activities, pictured in figure 10-35 (the first two, whose names begin with I). The two arcs transform raffinate in Texas for 1st time period and 2nd time period, respectively.

```
ANALYZE ... GR LOAD
ANALYZE ... GR COPY
ANALYZE ... GR SET * //TIME
                         9 columns deleted from submatrix
ANALYZE ... GR D R //S
                                          ROW DIGRAPH
                                 Flow for MT=unleaded regular, R=Texas
  1st time period --->2nd time period
  2nd time period --->3rd time period
```

Figure 10-39. Setting and Displaying Row Digraph for Transformations of Time

The general rule in using set attribution to delineate transformations is as follows. For each activity in the submatrix, its input/output variation is determined by the domain sets of the equations it intersects. That is, suppose set A is attributed with some property, like form, and suppose an activity has a negative entry in equation B(a,...), where set A is in the domain of equation B and a is a member of A. Further suppose the activity has a positive entry in equation C(a′,...), where a′ is a member of set A′ that has the same attribute as A (it can even be that A=A′). Then, the activity is a transformation of form (or whatever property has been attributed to sets A and A′). If that is the attribute specified, the activity remains in the submatrix. Otherwise, if the activity does not transform any set with the specified attribute, it is deleted from the submatrix before the graph is set.

For example, when we examine a transportation activity, say T(MT, SR, DR), where its domain is MT×SR×DR, we see that it has a negative entry in a row, say S(MT, SR), where the supply region, a member of SR, is the same in T as in S. The activity also has a positive entry in a row, say D(MT, DR), where the demand region, a member of DR, is the same in T as in D. Then, if sets SR and DR have both been attributed, say as place, the I/O structure of T is that it has a member of MT×SR as an input and a member of MT×DR as an output, where the member of MT is the same. Thus, T is a transformation of place since the variation is from SR to DR, both of which are sets attributed with place.

It is the domains of the equations that determine the set variations of an activity. How the attributed sets vary determine the type of transformation represented by the activity. In our examples we have illustrated activities that transform only one attribute (form, place or time). It is possible for an activity to transform more than one attribute. For example, an activity can have ore in Pittsburgh as an input by having a negative entry in an equation that balances material×region (domain = MT×R, say). The same activity can have steel in Philadelphia as an output in another equation that also balances material×region. If material is attributed form and region is attributed place, the activity is a transformation of both form and place. We could even add time by having a lag between the input and the output.

Further general discussion is beyond the scope of this User's Guide, but you can modify the syntax files (or use MODLER) to experiment with this concept.

10.5 Schema Equations.

Figure 10-40 shows a display option for the SCHEMA command. The equations can be displayed instead of, or in addition to, the schema table (to follow along, read in WOODNET and its syntax file).

```
ANALYZE ...SCHEMA DISPLAY EQUATION
  S(MT,SR) =    S(MT,SR) - Sum[DR | T(MT,SR,DR)] = 0
  D(MT,DR) =    Sum[SR | T(MT,SR,DR)] - D(MT,DR) = 0
  COST     =    Sum[MT,SR | [10/60] S(MT,SR)]
             + Sum[MT,SR,DR | [5/20] T(MT,SR,DR)] ...MIN

                          0 <= S(MT,SR)    <= [5/*]
                          0 <= T(MT,SR,DR)
                   [5/100]  = D(MT,DR)     (Fixed Levels)

  The notation [a/b] means the coefficients, which can depend upon
  the domain member, is in the range a to b (see SCHEMA.DOC).
```

Figure 10-40. Equations of Schema for WOODNET

The keyword Sum denotes a summation. For example, consider the supply equation,

```
S(MT,SR) =   S(MT,SR) - Sum[DR | T(MT,SR,DR)] = 0
```

where the summation is over the set of demand regions, DR.

Algebraically, this equation is equivalent to the constraint:

$$S_{mi} - \Sigma_{j \in DR} T_{mij} = 0 \text{ for all } m \in MT \text{ and } i \in SR,$$

where the domain of the summation is inferred from the fact that set DR is in the domain of the activity class (T), but it is not in the domain of the equation.

As indicated by the footnote, ranges of coefficients are indicated by bracketing the range specification that appears in the schema table. On-line help is available with SCHEMA.DOC.

Chapter 11. Rulebase Development

...science as well as technology, will in the near and in the farther future increasingly turn from problems of intensity, substance and energy to problems of structure, organization, information, and control.

J. von Neumann

This chapter describes rulebase development – that is, using rule files to provide a form of intelligent analysis support. Chapter 6 gave the bare essentials of the rule file structure, the keywords and the commands. Here we expand their descriptions and discuss examples that came with ANALYZE (see, also, RULEBASE.DOC).

11.1 Basic concepts and tools.

A rule file is a way to use ANALYZE with responses to questions of analysis. It contains text, references to the LP information, and commands. Text appears in free format until LP information is needed. Then, there are two ways to insert information about the LP into the response that the user sees. The first way is with a simple keyword lookup. As shown in figures 6-6 through 6-8, the keywords can be character, integer, or real values. When referencing a keyword in the rule file, a percent sign (%) precedes its name. For example, consider the following *template*:

```
Problem %PROBLEM is %STATUS with objective to %OPT %OBJ .
```

Suppose the problem name is WOODNET, the solution status is OPTIMAL, the sense of optimization is MINIMIZE, and the name of the objective row is COST. Then, the user will see the sentence:

```
Problem WOODNET is OPTIMAL with objective to MINIMIZE COST.
```

Note that there is a space before the period at the end of the sentence in the template. This is so %OBJ can be parsed. The response formation shifts the period to give the correct punctuation for the sentence.

Text continues in this fashion, substituting values of keywords, until one of the following occurs: an ANALYZE command is specified; the TEXT command is specified; the INTERPRT command is specified; or, the EXIT command is specified. These commands are described in the next section. Until one of them is reached, the response text continues, using the screen width and length set by the user (with the SCREEN command). Blank lines are ignored, and all text is left-justified.

In addition to the system keywords, you can specify your own *parameters* (see the SET command in the next section). These are referenced in the text with a double percent sign (%%). Special parameters, named 1, 2, ..., 9, as in a DOS bat file, are passed from the INTERPRT command specification. For example, in Chapter 4 automatic price interpretation was illustrated with the INTERPRT PRICE command (we shall step through the PRICE.RUL rule file shortly). Before the specification, the user had to put the row into the submatrix (eg, with DISPLAY or SUBMAT). Alternatively, the user can specify INTERPRT PRICE DMOCH to apply this to row DMOCH. Upon instantiation of the rule file, parameter 1 is set to DMOCH. The user can specify up to nine parameters in this way, and they are referenced in the rule file by %%1, %%2, ..., %%9. If the user does not specify a row, parameter 1 is the null string.

11.2 Rule file commands.

Figure 6-9 lists the commands recognized in a rule file. Each command begins on a separate line and has a dollar sign ($) in column 1. This section describes their functions and syntax more fully. The remaining sections illustrate their use.

> Each command must be fully specified; no abbreviation of the command is permitted.

The commands are given in alphabetical order (as in figure 6-9), beginning with their syntax. Notes and examples follow each.

$ANALYZE *command spec*

The *command spec* can be any ANALYZE command, except the INTERPRT command. Since ANALYZE is written is FORTRAN/77, it is not recursive, so the INTERPRT command has its own specification from a rule file. All other ANALYZE commands, both session control and procedural, can be executed, and the output is the same as if the user executed them directly. It is unwise, however, to use a command involving I/O, such as the OUTPUT command. The user might have set an output file for the rule response, and if the rule file specifies the OUTPUT command, it will close the user's file. Text is cleared before executing any ANALYZE command.

Examples:

`$ANALYZE EXPLAIN ROW`

.......Explains rows in the submatrix

`$ANALYZE SUB R %OBJ`

.......Sets submatrix to have the objective row

`$ANALYZE D C %%MYCOL`

.......Displays column(s) whose name mask is the value of the parameter, MYCOL

$CALC *parameter* = *expression*

expression := *term* | *term* ± *expression*

term := *numeric* [* *numeric*]

numeric is a numeric constant, or a reference to a keyword or parameter whose value is numeric.

The first term can be the parameter on the left of the equals sign without the double percent sign.

Examples:

```
$CALC N = N + 1
```
.......Increments the value of parameter N.
```
$CALC N = %%A + %NCOLS
```
.......Sets N equal to the sum of the value of the parameter, A, and the keyword NCOLS (number of columns in the submatrix).
```
$CALC N = %%A + %%B - %%C
```
.......Sets N equal to the sum of parameters A and B minus the value of parameter C
```
$CALC N = 2 * %NROWS
```
.......Sets N equal to 2 times the number of rows in the submatrix.

$DEBUG [*comment*]

Pauses in the rule file and shows the comment on the screen. Used as a debugging aid.

$ENDLOOP

Specifics end of a loop (see LOOP command). It also is used for branching to the statement immediately following the end of the loop (which could be a NEXT command).

$ENTITY *member* [*set*]

Translates the string *member* in the *set*. If set is not specified, it translates *member* from all sets in which the string appears. If *set* is specified, and the set does not exist (or if the syntax is not resident), an error message is issued. If *member* is not in the set, no translation or error message is given.

Examples:

`$ENTITY %%MT MT`
.......Translates the (string) value of parameter MT as a member of set MT
`$ENTITY 3a`
.......Translates 3a in all sets that contain 3a as a member

$EXIT

Terminates rule file (after clearing remaining text).

$GOTO *label* | TOP

Unconditional branch to line immediately following *label*, or to the top of the rule file. A label appears on a separate line with a colon (:) in column 1. The label can be 1 to 8 characters. Outside a loop, the branch to a label is only forward. Inside a loop, the label can occur anywhere. The search is forward until the end of the loop is reached, then it searches from the top of the loop. If the label is not found in the loop, the loop is ended, and search proceeds after the loop through the remainder of the rule file. If the label is not found, an error message is given.

Examples:

`$GOTO A`
.......Branches to the statement following
:A
`$GOTO TOP`
.......Branches to top (rewinds rule file)

$IF *condition* THEN *command*

The *command* is any rule file command. The *condition* is a simple comparison, *a rel b*, where *a* and *b* are any strings, including keyword and parameter references; and, *rel* is one of the 6 relations: <, < =, =, < >, > =, >. If the *condition* is true, the *command* is executed; otherwise, processing continues with the next line.

Examples:

`$IF %%N = 1 THEN GOTO A`
.......If the value of the parameter, N, equals 1, control branches to the statement following the label A. Otherwise, processing continues with the next line.
`$IF %NROWS > 1 THEN SKIP 2`
.......If the number of rows in the submatrix is greater than 1, the next 2 lines are skipped.

```
$IF %STATUS <> OPTIMAL THEN EXIT
```
.......If the problem status is not OPTIMAL, the rule file is terminated.

$INTERPRT *filespec*

The rule file, *filespec*, is instantiated, and the present rule file is terminated. The RULEBASE type in _SETUP applies, but the rule file need not be in RULBASE.RUL (it can be *hidden*).

Examples (assume default prefix is blank and suffix is .RUL):

```
$INTERPRT PRICED
```
.......Instantiates rule file, PRICED.RUL
```
$INTERPRT PRICE%%TYPE
```
.......Instantiates rule file, PRICEt, where t is the value of the parameter, TYPE.

$LOOKUP *key name* [,*name*]

Looks up the *key* value for *name*(s) and puts result into the value of the keyword, VLOOK. The *key* and *name*(s) are as follows.

{X D C XL XU} *column*
{Y P S YL YU} *row*
NZ *row*, *column*
BLOCK {ROW COL} *block_name*

The keys for a column name are level (X), reduced cost (D), objective coefficient (C), lower bound (XL), and upper bound (XU). The keys for a row name are level (Y), price (P), slack (S), lower bound (YL), and upper bound (YU). The key NZ is the coefficient value in *row*, *column*. The BLOCK {ROW COL} key returns the number of rows or columns, respectively, in the block whose name is *block_name*. This last is usually used to determine whether a block that was created by some procedure (such as ANALYZE BLOCK NETFORM) is null.

Examples:

```
$LOOKUP X %%MYCOL
```
.......Puts into VLOOK the level of the column whose name is the value of the parameter, MYCOL.
```
$LOOKUP P B%%MT%%R
```
.......Puts into VLOOK the price of the row whose name is Bmr, where m is the (string) value of the parameter MT, and r is the (string) value of the parameter R.
```
$LOOKUP NZ %ROW,%COLUMN
```

.......Puts into VLOOK the coefficient of the current row and column.

```
$LOOKUP BLOCK COL NET.PROD
```

.......Puts into VLOOK the number of columns in the column block whose name is NET.PROD (viz., this is the block of production activities created by BLOCK NETFORM)

$LOOP

This begins a loop over rows and/or columns. The loop ends with either the $ENDLOOP or $NEXT command. Examples of loops are given in remaining sections.

$NEXT [ROW] [COLUMN]

This advances the submatrix row and/or column pointers, %ROW and %COLUMN, respectively. Then, it branches to the statement immediately following the LOOP command. If a NEXT statement is reached outside a loop, an error message is given. If neither ROW nor COLUMN is specified, both are assumed. After the last row and/or column, the loop ends.

Examples:

```
$NEXT R
```

.......Advances to next row in submatrix.

```
$NEXT C
```

.......Advances to next column in submatrix.

```
$NEXT R, C
```

.......Advances to next row and column in submatrix.

```
$NEXT
```

.......Same as above.

$POP {1 2} *parameter*

Pops last entry of stack number (1 or 2) and puts value into *parameter.* If the stack is empty, the value of parameter is null.

Example:

```
$POP 1 COL
```

.......Pops stack 1 and puts value into parameter, COL.

$PUSH {1 2} *string*

Pushes *string* onto stack number (1 or 2). The string can be a literal or a reference to a keyword or parameter. A push puts the entry at the top of the stack.

Examples:

`$PUSH 1 %%COL`

.......Pushes the value of parameter COL onto stack 1.

`$PUSH 2 %COLUMN`

.......Pushes the name of the current column onto stack 2.

$QUEUE {1 2} *string*

Queues *string* onto stack number (1 or 2). The string can be a literal or a reference to a keyword or parameter. A queue puts the entry at the bottom of the stack.

Examples:

`$QUEUE 1 %%COL`

.......Queues the value of parameter COL onto stack 1.

`$QUEUE 2 %COLUMN`

.......Queues the name of the current column onto stack 2.

$SET *parameter = string*

This defines a new parameter or changes the value of an existing parameter. The *string* can be literal or a reference to a keyword or parameter.

Examples:

`$SET A =`

.......Sets the value of parameter A to be null

`$SET N = 0`

.......Sets the value of parameter N to be 0

`$SET A = N`

.......Sets the value of parameter A to be N

`$SET A = %%N`

.......Sets the value of parameter A to the value of parameter N

`$SET N = %NCOLS`

.......Sets the value of parameter N to the number of columns in the submatrix.

$SET N

.......Clears parameter N (note no equal sign following N).

$SKIP *number* | LOOP | ENDLOOP

This skips *number* of lines (not counting blanks or comments), or to the end of the loop (NEXT statement), or to the statement immediately following the end of the loop (terminates the loop).

Examples:

`$SKIP 2`
.......Skips 2 lines
`$SKIP LOOP`
.......Skips to the end of the loop (to the NEXT statement)
`$SKIP ENDLOOP`
.......Ends the loop

$STACK [{1 2}]

Clears stack(s). If no number is specified, both stacks are cleared.

Examples:

`$STACK`
.......Clears stacks 1 and 2
`$STACK 1`
.......Clears stack 1

$TEXT [*number* [MARGIN=*margin* [*indent*]] | CENTER]

Clears text and skips *number* − 1 lines. If MARGIN is specified, which must be after *number*, the left margin is set to *margin*. In that case, the second value, if present, sets the indent. Alternatively, the CENTER option centers whatever is in the text buffer. This should be one line that is within the screen width.

Examples:

`$TEXT`
.......Clears text (if there is no text, a blank line is generated).
`$TEXT 1`
.......Same as above.
`$TEXT 2`
.......Clears text and generates 1 blank line.
`$TEXT 1 MARGIN=5`
.......Clears text and sets left margin to 5, which applies to subsequent text.
$TEXT CENTER
.......Centers text and clears it (margin and indent settings are ignored).

11.3 Simple examples.

In this section we describe some of the simple rule files that came with ANALYZE. These do not contain reasoning, as the examples in the subsequent sections, but they demonstrate some of the commands and keywords. They also illustrate some of the ways short reports can be prepared without LP jargon.

We begin with the rule file, SOLSTATS.RUL, which came with ANALYZE. As shown in figure 1, the first command is to put all rows and columns into the submatrix. Then, free rows are deleted by the second command. The subsequent two SET commands store the numbers of non-free rows and columns into parameters, NR and NC, respectively.

The first text appears on line 5, where the response text is formed after replacing the keyword references with their values. First, %NROWS is replaced by the number of rows in the submatrix (which equals the number of constraint equations, having removed the free rows). Second, %NCOLS is replaced by the number of columns in the submatrix. Third, %PROBLEM is replaced by the problem name.

The text is cleared before executing the subsequent ANALYZE command, which sets the submatrix to all rows whose dual price is not zero (since C = 0 for rows, except the objective). The next ANALYZE command removes the objective row from the submatrix, so what remains are the active rows – that is, constraints with nonzero prices.

The TALLY command is then used to display the active row classes and the number in them. Then, the activities with positive level are tallied (where level less than .01 is presumed to be zero). Finally, degeneracy statistics are determined and reported.

```
$ANALYZE SUB *
$ANALYZE SUB D R * L=-* U=*
$SET NR = %NROWS
$SET NC = %NCOLS
  The %NROWS rows (constraint equations) and %NCOLS columns
  (activities) in %PROBLEM have the following solution distributions.
$ANALYZE SUB R * P NE C
$ANALYZE SUB D R %OBJ
  Active rows (nonzero prices):
$ANALYZE TALLY ? R
  Activities with positive levels:
$ANALYZE SUBMAT COL * X = .01/*
$ANALYZE TALLY ? C
  Primal degeneracies:
$ANALYZE SUB COL * S=B X EQ L
$ANALYZE SUB A C * S=B X EQ U
$ANALYZE TALLY ? C
$ANALYZE SUB ROW * S=B Y EQ L
$ANALYZE SUB ROW * S=B Y EQ U
$ANALYZE TALLY ? R
  Dual degeneracies:
$ANALYZE SUB ROW * S=LU P=0
$ANALYZE TALLY ? R
$ANALYZE SUB COL * S=LU D=0
$ANALYZE TALLY ? C
$EXIT
```

Figure 1. Rule File, SOLSTATS.RUL

Figure 2 shows the first portion of the rule file, STATS.RUL, which came with ANALYZE. The first line forms a header, where the problem name is substituted for the keyword reference, %PROBLEM. If, for example, the problem is WOODNET, the user will see:

```
Statistical Summary of WOODNET
```

The TEXT command causes this to be generated, centering the text (by //CENTER) and clearing the text buffer for what follows.

Then, the numbers of null rows and columns are obtained from ANALYZE SUBMAT command specifications and put into parameters, NR0 and NC0, respectively.

```
     Statistical Summary of %PROBLEM
$TEXT CENTER
$ANALYZE SUB ROW * N=0
$SET NR0 = %NROWS
$ANALYZE SUB COL * N=0
$SET NC0 = %NCOLS
$ANALYZE SUBMAT *
$SET NR = %NROWS
$SET NC = %NCOLS
$SET NZ = %NONZERO
$IF %%NR0 = 0 THEN GOTO R0END
$CALC NR = NR + %%NR0
$TEXT 1 MARGIN=5
 The following rows are null.
$ANALYZE DISP R * N=0 //
:R0END
$IF %%NC0 = 0 THEN GOTO C0END
$CALC NC = NC + %%NC0
$TEXT 1 MARGIN=5
 The following columns are null.
$ANALYZE DISP C * N=0 //
:C0END
```

Figure 2. First Portion of Rule File, STATS.RUL

The next line executes the ANALYZE command, SUBMAT *, which puts all rows and columns into the submatrix. Then, parameters NR, NC and NZ are defined with the SET command to equal the number of rows, columns and nonzeroes, respectively.

Line 11, the first IF statement, skips to label R0END if there are no free rows – that is, if parameter NR0 equals 0. Otherwise, the null rows are displayed. The same logic is used to display null columns, if any. This brings us to the statement label C0END.

The next portion, shown in figure 3, begins a series of commands to gather statistics to be reported. At the TEXT command, we have set the following parameters.

NR ....Number of rows
NC...........Number of columns
NZ...........Number of nonzeroes
NEQ ....Number of equations

NFR........Number of free rows
NLU........Number of inequality rows (including ranges)
NFX........Number of fixed columns
NV ....Number of "vanilla" columns (ie, simply non-negative)
NBD ....Number of bounded columns (nonzero lower bound and/or finite upper bound, but not fixed)

```
$ANALYZE SUBMAT R * L EQ U
*                    :..........Equations
$SET NEQ = %NROWS
$ANALYZE SUBMAT CO *
$ANALYZE SUBMAT R * L=-*  U=*
*                    :..........Free rows
$SET NFR = %NROWS
$SET NLU = %%NR
$CALC NLU = NLU - %%NFR - %%NEQ
*     :.........L-rows, G-rows and ranges
$ANALYZE SUBMAT R *
$ANALYZE SUBMAT CO * L EQ U
*                     :........Fixed columns
$SET NFX = %NCOLS
$ANALYZE SUBMAT R *
$ANALYZE SUBMAT CO * L=0 U=*
*                     :........Vanilla
$SET NV = %NCOLS
$SET NBD = %%NC
$CALC NBD = NBD - %%NV - %%NFX
*     :........Bounded, but not fixed
$TEXT 1 MARGIN=5
                  Dimensions
$TEXT CENTER
                  ~~~~~~~~~~
$TEXT CENTER
   %%NR Rows.......... %%NEQ Equations, %%NLU Inequalities, %%NFR Free
$TEXT
   %%NC Columns....... %%NV Vanilla, %%NBD Bounded, %%NFX Fixed
$TEXT
   %%NZ Nonzeroes..... %NVALUES Distinct, %NONES ones, Density = %VDENSTY
$TEXT
```

Figure 3. Second Portion of Rule File, STATS.RUL

The TEXT command not only clears the text buffer, but also it defines the left margin to be 5, which applies to subsequent output. Then, with next two TEXT commands, the user sees:

```
Dimensions
~~~~~~~~~~
```

which is centered, as specified.

Before continuing, you might read an LP into memory, such as WOODNET, and issue the command INTERP STATS, to see the result. Then, do this for another LP, such as FOREST.

Then, the rule file proceeds to use the parameters plus keywords to report the problem statistics. Each statistic is the value of either a parameter or the value of a keyword. The keywords used to produce the statistics are, in order of appearance, the number of distinct values (NVALUES), the number of ones (NONES), and the density (VDENSTY).

Figure 4 shows the third portion of this rule file, which reports on how much of the linear program is a netform. The first three ANALYZE commands erase whatever blocks may have been formed and puts all rows and columns into the submatrix. Then, $ANALYZE BLOCK NET executes the ANALYZE command to find a maximal embedded netform. This results in column blocks, named NET.PROD, NET.TRAN and NET.CONS (see chapter 5). If the LP has any activity with a single nonzero in non-free rows, it is put into NET.PROD if that nonzero is positive and into NET.CONS if it is negative. The block named NET.TRAN contains activities with exactly one positive nonzero and one negative nonzero, both in non-free rows (signs are treated as reversed for L-rows).

The three LOOKUP commands set the numbers of production, transfer, and consumption activities into parameters, #PROD, #TRAN, and #CONS, respectively. For example, consider the command $LOOKUP BLOCK COL NET.PROD. The number of columns in this block is put into the value of keyword VLOOK. Then, the command $SET #PROD = %VLOOK sets #PROD to this number.

The submatrix is then cleared to copy each of the blocks to it (each BLOCK COPY merges the column blocks with the submatrix). The IF ... THEN statements perform the block copy only if the block is not null (attempt to copy a null block causes an error message).

```
*  Get Embedded Netform
$ANALYZE BLOCK R E *
$ANALYZE BLOCK C E *
$ANALYZE SUBMAT  *
$ANALYZE BLOCK NET

$LOOKUP BLOCK COL NET.PROD
$SET #PROD = %VLOOK
$LOOKUP BLOCK COL NET.TRAN
$SET #TRAN = %VLOOK
$LOOKUP BLOCK COL NET.CONS
$SET #CONS = %VLOOK

$ANALYZE SUBMAT CLEAR
* Copy Netform to ANALYZE SUBMATrix
$IF %%#PROD > 0 THEN ANALYZE BLOCK COL COPY FROM NET.PROD
$IF %%#CONS > 0 THEN ANALYZE BLOCK COL COPY FROM NET.CONS
$IF %%#TRAN > 0 THEN ANALYZE BLOCK COL COPY FROM NET.TRAN
$ANALYZE SUBMAT  R *
$ANALYZE SUBMAT SET
$TEXT
  Embedded Netform is %NROWS x %NCOLS :
$TEXT 1 MARGIN=2
...%%#PROD Production, %%#CONS Consumption, %%#TRAN Transfer activities.
$TEXT
```

Figure 4. Third Portion of Rule File, STATS.RUL

After the block copies are finished, all rows are put into the submatrix, and the submatrix is set – that is, null rows and columns are removed with the ANALYZE SUBMAT SET command. The TEXT command produces a blank line since the text had previously been cleared.

The text now specifies:

```
Embedded Netform is %NROWS x %NCOLS :
```

ANALYZE substitutes the numbers of rows and columns, which are the respective values of the keywords, NROWS and NCOLS. These numbers represent the size of the maximal embedded netform. The final colon is shifted left in forming the response text because it is punctuation; it has a blank in the rule file to separate it from %NCOLS.

The TEXT command again clears the text, so the above appears on a separate line. The subsequent line of text gives the numbers of production, transfer and consumption activities. Finally, before beginning the next portion, the TEXT command again clears the response text.

In figure 5, we see a conditional branch. If the status of the solution is OPTIMAL, processing continues with the TEXT command. Otherwise, the objective is given, then the rule file terminates with the EXIT command. Let us pursue the case of an optimal solution.

```
$IF %STATUS = OPTIMAL THEN SKIP 2
    Objective:  %OPT %OBJ ...Resident Status = %STATUS .
$EXIT
$TEXT 1 MARGIN=1
                Solution Statistics
$TEXT CENTER
                ~~~~~~~~~~~~~~~~~~~
$TEXT CENTER
$ANALYZE SUBMAT R %OBJ
    Objective:     %OPT(1:3) %OBJ = %VROWY
$ANALYZE SUBMAT R * S B Y EQ U
$SET NR = %NROWS
$ANALYZE SUBMAT R * S B Y EQ L
$CALC NR = NR + %NROWS
$ANALYZE SUBMAT CO * S B X EQ L
$SET NC = %NCOLS
$ANALYZE SUBMAT CO * S B X EQ U
$CALC NC = NC + %NCOLS
    Primal Degeneracies:  Rows = %%NR   Columns = %%NC
$ANALYZE SUBMAT R  * S LU P=0
$ANALYZE SUBMAT CO * S LU D=0
    Dual Degeneracies:    Rows = %NROWS   Columns = %NCOLS
$ANALYZE SUBMAT R  * S=LU
$ANALYZE SUBMAT CO * S=BI
    The basis has %NROWS binding rows (and %NCOLS basic structurals)
    with %NONZERO nonzeroes.
$EXIT
```

Figure 5. Final Portion of Rule File, STATS.RUL

The "Solution Statistics" begins with a header and the objective value. Consider the text:

```
Objective:  %OPT(1:3) %OBJ = %VROWY
```

The keyword reference, %OPT(1:3), substitutes the first three characters of the value of OPT. If the sense of optimization is MINIMIZE, the first portion of this text appears as: `Objective:` MIN (if it is MAXIMIZE, MAX replaces MIN). Then, the name of the objective is put into the response text (i.e., substituting the value of the keyword, OBJ), followed by its level (i.e., value of keyword VROWY, where the current row was set to the objective row by the previous command, $ANALYZE SUBMAT R %OBJ.

The remaining commands set degeneracy statistics and report them, terminating after giving the basis statistics.

Try this with several linear programs (see DIRMATS for a directory) to follow the instantiation of these two rule files. For the remaining examples, you will also need to read in the syntax files.

11.4 Price interpretation.

In chapter 4 automatic interpretation of an optimal dual price was illustrated with WOODNET. We entered INTERPRT PRICE to instantiate a rule file named PRICE.RUL. Now we examine this rule file and explain its contents.

Figure 6 shows the first portion, which contains checks and simple, generic tests. The rule file uses the IF command to see if the user specified a row name, passed as parameter 1. If so, it is put into the submatrix with the ANALYZE SUBMAT command.

The next line, which has * in column 1, is merely a comment and is ignored. The next line uses the IF command to check that the submatrix contains a row. If so, the value of the keyword, NROWS, will be positive, in which case we skip to the line following the EXIT command. If not, the error message is given, and the rule file is terminated.

A similar check is made that there is exactly one row in the submatrix. Then, if that row is the objective, a special rule file is instantiated (which the user could have specified directly, as INTERPRT OBJECTIV). In this case, the instantiation of OBJECTIV.RUL takes control out of this rule file, never to return.

```
$IF @%%1 <> @ THEN ANALYZE SUBMAT ROW %%1
* CHECK THAT A ROW WAS SET
$IF %NROWS > 0 THEN SKIP 2
 No row specified (Use SUBMAT R name)
$EXIT
* CHECK THAT ONLY 1 ROW WAS SET
$IF %NROWS = 1 THEN SKIP 2
 More than 1 row was specified...Please INTERPRT 1 at a time
$EXIT
$IF %ROW = %OBJ THEN INTERPRT OBJECTIV
* ROW IS NOT OBJ...
$IF %STATUS <> OPTIMAL THEN GOTO RATE
$IF %SWSYN = F THEN GOTO RATE
(some comments appear)
$SET ROW = %ROW
$SET PRICE = %VROWP
* GIVE MEANING OF THE ROW TO START WITH
$ANALYZE EXPLAIN ROW
* SEE IF ROW IS SLACK
$IF %%PRICE <> 0 THEN SKIP 2
 This row is not binding, so its marginal price = 0.
$EXIT
```

Figure 6. First Portion of the Rule File, PRICE.RUL

Otherwise, we then check that the solution status is optimal with the command,

```
$IF %STATUS <> OPTIMAL THEN GOTO RATE
```

ANALYZE first substitutes the value of the keyword, STATUS, which can be OPTIMAL, INFEASIBLE, UNBOUNDED, or unknown. If it is not OPTIMAL, the condition succeeds, in which case processing continues at the statement following the label RATE (see below). If the solution status is OPTIMAL, we fall through to check if the syntax has been read into memory by the test,

```
$IF %SWSYN = F THEN GOTO RATE
```

ANALYZE first substitutes the value of the keyword, SWSYN. It is T (true) or F (false), according to whether there is syntax in memory. If not, the condition succeeds and we branch to RATE. Otherwise, if we fall through, we have the syntax in memory, which brings us to the settings of parameters ROW and PRICE.

The next response text is the result of the ANALYZE command, EXPLAIN ROW, which gives the syntactic translation of the row set by the user. If the syntax file has not been read into memory, an error message is given.

Now the first generic test is made, namely the case of a zero price. Regardless of which linear program is resident, a generic interpretation is given:

```
This row is not binding, so its marginal price = 0.
```

Otherwise, we continue processing with the second portion of PRICE.RUL, shown in figure 7. Recall, from chapter 4, that we began the analysis with putting into the submatrix all basic activities, then executing the TRACE command. This is what the next two commands do.

Next, the ANALYZE BLOCK command is used to save the submatrix because it can change with some commands to follow. Thus, note the row and column blocks are erased (in case there were any), then the submatrix rows and columns are copied to blocks, ROWS and COLS, respectively. The ANALYZE BLOCK SAVE then saves these blocks, which we can load later if necessary.

The second generic test is performed – that is, seeing if there is a basic activity with a nonzero coefficient in the row in question and in no other non-free row. The TRACE command does not include nonfree rows, except the objective, so the next command deletes the objective row from the submatrix before beginning the loop.

The first test in the loop is the command:

```
$IF %NZCSUB > 1 THEN SKIP LOOP
```

The keyword, NZCSUB, is the number of nonzeroes of the current submatrix column in the submatrix rows. If this is greater than 1, the SKIP LOOP branches to the NEXT C command, which goes to the next column in the submatrix. (Once all columns have been interrogated, the loop ends.)

```
* PUT BASIC COLUMNS INTO SUBMATRIX TO INITIALIZE TRACE
$ANALYZE SUBMAT CO * S=B
$ANALYZE TRACE
* SAVE SUBMATRIX
$ANALYZE BLOCK ROW ERASE *
$ANALYZE BLOCK COL ERASE *
$ANALYZE BLOCK ROW COPY TO ROWS
$ANALYZE BLOCK COL COPY TO COLS
$ANALYZE BLOCK SAVE
* SEE IF THERE IS A SINGLETON BASIC COLUMN IN THIS ROW
$ANALYZE SUBMAT D R %OBJ
$ANALYZE SUBMAT SET
$LOOP
$IF %NZCSUB > 1 THEN SKIP LOOP
$IF %VCOLD <> 0 THEN SKIP LOOP
$LOOKUP NZ %%ROW, %COLUMN
$IF %VLOOK = 0 THEN SKIP LOOP
* WE HAVE SINGLETON BASIC COLUMN IN THIS ROW...SIMPLE EXPLANATION
Its price of %%PRICE is the unit (input) %OBJ of activity %COLUMN, where
$ANALYZE EXPLAIN COL %COLUMN
$IF %VCOLC = %%PRICE THEN GOTO EXIT
$TEXT
Note:  the activity cost = %VCOLC, but each level of activity has
       %%VLOOK in row %%ROW .
$GOTO EXIT
$NEXT C
```

Figure 7. Second Portion of the Rule File, PRICE.RUL

If the condition fails, the current column has a single nonzero in the rows of the submatrix, and the next command checks if its reduced cost (%VCOLD) is nonzero. If so, it cannot provide the generic interpretation we seek, so we skip to the next column. If so, we use the LOOKUP command to see if its nonzero is in the row in question. Having set parameter ROW to this row (c.f., figure 6), the names are given by:

```
%%ROW, %COLUMN
  |     |
  |     └──► keyword value = name of current column
  └──► parameter value = name of row in question
```

The coefficient is put into the value of keyword VLOOK, which is then tested to see if it is zero. If so, we skip to the next column. If not, we have the generic interpretation we sought. The dual price of the row is simply C_j/A_{ij}, where C_j is the objective coefficient of this activity (column j) and A_{ij} is the column's coefficient in the row in question (i). The user sees the interpretation that begins with the clause:

```
Its price of %%PRICE is the unit (input) %OBJ of activity %COLUMN, where
```

after the parameter and keyword references are replaced by their values. The next line in the rule file completes the sentence with the syntactic explanation of the column.

If the objective coefficient equals the price, the interpretation is complete. Otherwise, the coefficient value (VLOOK) is not ±1, and the note is given. In either case, the unconditional branch, GOTO EXIT, terminates the rule file.

If no such column is found, processing continues after the loop with the next portion, shown in figure 8. This begins by restoring the submatrix (recall that the objective row was deleted). The ANALYZE SUBMAT SET command is necessary to have the correct submatrix statistics. Once this is done, the blocks are erased.

Having failed the generic tests, processing continues with rule files that are bound to the particular model. (The resident linear program is an instance of the model; what is presumed from this point is the model's syntax.) For price interpretation, only WOODNET and BLENDRT models can continue. If the LP is some other model (eg., FOREST), processing continues by branching to RATE.

The processing following the RATE label is generic. Recall that we get there if any of the following conditions hold: solution is not optimal, there is no syntax in memory, or the model is not one of the designated ones (WOODNET or BLENDRT). In this case, a short preamble is given before showing the rates of substitution with the //P option (try it).

In the case of WOODNET, the first character of the row name must be D or S (demand or supply, respectively). In that case, the command INTERPRT PRICE%%ROW(1:1), instantiates rule file PRICED.RUL or PRICES.RUL, respectively. In the case of BLENDRT, the only row class is the balance row, which begins with B, so PRICEB.RUL is instantiated. The PRICE.RUL file is closed.

```
$ANALYZE BLOCK LOAD
$ANALYZE BLOCK ROW COPY FROM ROWS
$ANALYZE BLOCK COL COPY FROM COLS
$ANALYZE SUBMAT SET
$ANALYZE BLOCK ROW ERASE *
$ANALYZE BLOCK COL ERASE *
* NOW THE INTERPRETATION IS DEPENDENT ON THE ROW CLASS
* (AND HENCE THE MODEL)
$IF %PROBLEM = WOODNET THEN GOTO OK
$IF %PROBLEM = BLENDRT THEN GOTO OK
$GOTO RATE

:OK
 I shall try to interpret its price of $%%PRICE .
$INTERPRT PRICE%%ROW(1:1)
:EXIT
$ANALYZE BLOCK ROW ERASE *
$ANALYZE BLOCK COL ERASE *
$EXIT
:RATE
* THIS IS GENERIC INTERPRETATION...USING RATEOF
$SET V = %VROWUP
$IF %%V = * THEN SET V = %VROWLO
Note the following effects (ie, rates of substitution) a change in %ROW
has on the basic activity levels.  The "Net Rate" is the amount each basic
level must change times the %OBJ of the basic activity.  The sum of these
effects is the price of %ROW, which is the net rate at which %OBJ changes
with respect to a slight change in %ROW, which is currently = %%V .
$ANALYZE _PAUSE
$ANALYZE RATEOF ROW %ROW //P
$EXIT
```

Figure 8. Last Portion of the Rule File, PRICE.RUL

Let us continue with a demand row interpretation in WOODNET. The first portion of rule file PRICED.RUL is shown in figure 9.

The first two commands set the material and location according to the demand row's syntax. Parameter MT is set to characters 2 and 3 in the row's name by the reference, %%ROW(2:3). Similarly, parameter LO is set to characters 4 and

5 in the row's name by the reference, %%ROW(4:5). Then, parameters NS and DEGEN are set to 0.

A loop begins to calculate the number of suppliers and degeneracies. These values are put into parameters NS and DEGEN, respectively.

```
* ASSUMES VARIABLES:  ROW, PRICE AND SYNTAX D(MT,LO,...)
$SET MT = %%ROW(2:3)
$SET LO = %%ROW(4:5)
$SET NS = 0
$SET DEGEN = 0
* LOOP TO GET NUMBER OF SUPPLIERS
$LOOP
$IF %COLUMN(1:3) <> T%%MT THEN SKIP LOOP
$CALC NS = NS + 1
$IF %VCOLX = 0 THEN CALC DEGEN = DEGEN + 1
$NEXT C

$SET RENT = 0
$SET SOURCE
 The consumers in
$ENTITY %%LO DR
receive
$IF %%NS > 1 THEN SKIP 1
all of their
$ENTITY %%MT MT
from
$SET CONJUN=.
$SET PLURAL =
$SET NSPOS = %%NS
$CALC NSPOS = NSPOS - %%DEGEN
$IF %%NSPOS = 1 THEN SKIP 3
$SET PLURAL=s
%%NSPOS suppliers -- namely, from
$SET CONJUN=; and from
$SET SOURCE=@
```

Figure 9. First Portion of the Rule File, PRICED.RUL

The first test in the loop sees if the submatrix column is a transportation activity of the material with the condition. The string %COLUMN(1:3) is the first three characters of the column name. The string T%%MT is Txx, where xx is the value of parameter MT – that is, the material for the row in question. If these strings are not equal, we skip to the next column (ie., to the NEXT C command).

If these strings are equal, the column transports the material from some source, so the CALC command is used to increment the number of suppliers (parameter NS). Further, if the level of the transportation activity is zero, this is a degeneracy, so parameter DEGEN is incremented as well.

Once the loop ends, parameters RENT and SOURCE are initialized to 0 and null, respectively. The next three lines begin a sentence:

```
 The consumers in
$ENTITY %%LO DR
receive
```

The first of these lines form the literal text, 'The consumers in'. Then, the ENTITY command looks up the location (the value of parameter LO) in the set of demand regions (DR). This translation is inserted. The text continues with the literal, 'receive'. For example, suppose the row is DMOCH. Then, the value of %%LO is CH, and set DR contains member CH with the meaning: Chicago. The text is then:

```
The consumers in Chicago receive
```

The continuation of this text depends on whether there is only one source or many sources. If there is only one source (i.e., NS = 1), the text adds the phrase, 'all of their'. In any case, the next text added is from the translation of the material. This is obtained from the ENTITY command, which looks up %%MT in the set of materials (MT). Then, the literal, 'from', is added.

For example, if there is one source, the text at this point is:

```
The consumers in Chicago receive all of their mahogany from
```

(where 'mahogany' is the meaning of the material, MO). On the other hand, if there is more than one source, the text is:

```
The consumers in Chicago receive mahogany from
```

Before continuing with the response text, parameters CONJUN and PLURAL are initialized to . and null, respectively. This is so the resulting sentence will be in proper English. Then, parameter NSPOS is defined to be the number of actual sources – that is, activities with positive levels. This is done first with the SET command to define the parameter and set its value equal to the number of sources, then with the CALC command, which subtracts the number of degenerate sources (DEGEN).

If the number of actual sources is greater than 1, we re-set parameter PLURAL to 's', and we add text that gives the number of suppliers. Further, we change the conjunction (CONJUN) to be a semi-colon plus the phrase, 'and from'. In either case the value of parameter SOURCE is set to @ (this is an arbitrary value; it just needs to be non-null).

This brings us to figure 10. The loop is over the same columns, so the first two tests are the same as before (c.f., figure 9). Now parameter NSPOS is decremented with the CALC command, and the objective coefficient (ie., cost) of the column is set to the value of parameter, DP. (At the end DP will equal the delivered price of the material.)

The supply cost of the activity S%%MT%COLUMN(4:5)1 is set equal to the value of parameter SCOST from the LOOKUP. This is according to the WOODNET syntax. Given we have a transportation activity with positive level, the supply activity must also have positive level and hence contribute to the delivered cost. The name of the supply activity has syntax Smtlo1, where mt is the material (= %%MT) and lo is the location (= %COLUMN(4:5)). Its reduced cost is obtained from the LOOKUP and added to the rent. If this supply activity is basic, its reduced cost must be zero, but if it is at its upper bound, the reduced cost can be negative. The subtraction of %VLOOK from 0 calculates the value of parameter RENT as a non-negative quantity.

The parameter ASUP is then set equal to the source location string in the name of the column, and it is translated with the ENTITY command. At this point, the text is:

```
The consumers in Chicago receive [all of their] mahogany from
supplier with delivered cost = %%DP
```

where 'all of their' appears if there is more than one supplier; *supplier* is the translation from the ENTITY command; and, %%DP is replaced by the delivered cost, which was calculated as the sum of the supply activity cost and the transportation cost. For example, if the transportation activity is TMOLACH, *supplier* is Los Angeles, obtained from translating member LA in set SR.

If the value of parameter RENT is positive, a parenthetical phrase is added to the text; otherwise, this parenthetical is skipped. In either case, the conjunction is added to the text. Recall that if there is only one supplier, the conjunction is a period; otherwise, it is a semi-colon with text, 'and from'. In the case of more than one supplier, once we reach the last supplier, the value of NSPOS becomes 1, so the value of CONJUN changes to be a period.

```
* LOOP OVER SUBMATRIX TO GET DELIVERY PATHS AND COSTS
$LOOP
$IF %COLUMN(1:3) <> T%%MT THEN SKIP LOOP
$IF %VCOLX = 0 THEN SKIP LOOP
$CALC NSPOS = NSPOS - 1
$SET DP = %VCOLC
$LOOKUP C S%%MT%COLUMN(4:5)1
$SET SCOST = %VLOOK
$CALC DP = DP + %VLOOK
$LOOKUP D S%%MT%COLUMN(4:5)1
$CALC RENT = 0 - %VLOOK
$SET ASUP = %COLUMN(4:5)
$ENTITY %%ASUP SR
with delivered cost = %%DP
$IF %%RENT = 0 THEN SKIP 1
(this supplier, however, has an economic rent of %%RENT )
%%CONJUN

$IF %%NSPOS = 1 THEN SET CONJUN =.
$IF %%DP <> %%PRICE THEN SKIP LOOP
$SET SOURCE = %COLUMN(4:5)
$SET DTCOST = %VCOLC
$SET DSCOST = %%SCOST
$SET DSRENT = %%RENT
$NEXT C
```

Figure 10. Second Portion of the Rule File, PRICED.RUL

Then, processing continues with a test of whether the delivered price is not equal to the marginal price in question. If this condition is true, control skips to the next column in the loop. If, however, the delivered cost equals the marginal price, the source, transportation cost, supply cost, and rent are put into the values of parameters SOURCE, DTCOST, DSCOST and DSRENT, respectively.

This brings us to figure 11. If no simple path was found, the value of SOURCE never changed from its initial value, @. In that case, we branch to the text following label DC. Otherwise, advice is given to interpret the supply prices (PLURAL is 's' with more than one supplier). The last supply row is mentioned in the final phrase.

```
$IF %%SOURCE <> @ THEN GOTO DC
 The marginal price (%%PRICE ) = delivered cost + rent.
 You might try interpreting the supply price%%PLURAL
 ...see row S%%MT%%ASUP .
$EXIT
:DC
 Thus, marginal price = delivered cost = %%PRICE
 (from
$ENTITY %%SOURCE SR
 ) = price of %%ROW
 = Supply cost (%%DSCOST ) + Transportation cost (%%DTCOST )
$IF %%DSRENT <> 0 THEN SKIP 2
.
$EXIT
 + Supply rent (%%DSRENT ).
$EXIT
```

Figure 11. Last Portion of the Rule File, PRICED.RUL

If label DC is reached, the text begins a sentence with the marginal price decomposed into supply cost + transportation cost. If there is a positive rent, this is mentioned in the final phrase.

We have gone through PRICE.RUL and PRICED.RUL in detail. You should print these and try them in a session, beginning with EXEC WOODNET (to read both the packed LP file and the syntax file into memory). The logic in PRICES.RUL is similar. Then, you should interpret several different balance rows in BLENDRT (the packed file was created by ATEST.EXC after installation) and follow the formation of the response in PRICEB.RUL because it is different from PRICED and PRICES in both the reasoning and in the form of the response.

11.5 Testing new rule files.

In this section we illustrate a technique for testing new rule files. It can happen that you have several different rules in mind for the same interpretation, and you want to test them with different cases to see how each rule file variation behaves. One way to do this is to change RULBASE.RUL to have each of these rule files in the list. Another way is to have a single option, which we shall call TEST, which instantiates the particular rule of interest from a string variable.

This is included in TEST.RUL, which tests if string variable TEST has been set. If so, it instantiates it as a rule (otherwise, it conducts a test to check that the rulebase module works). We shall illustrate this with one case. Another case, which you might examine, is setting the string, TEST, equal to BLENDRT, which then instantiates rule file BLENDRT.RUL (begin by EXEC BLENDRT).

Figure 12 illustrates with a rule file that is instantiated from TEST.RUL if the string variable TEST is defined to be DISTRIB. There are three variations we want to explore, which pertain to showing the distribution of harvesting that meets demands in the FOREST model. The first command, therefore, tests that the resident problem is FOREST. If not, the response text says so and the rule file is terminated. If so, the processing branches to BEGIN.

```
$IF %PROBLEM = FOREST THEN GOTO BEGIN
Sorry, distribution analysis has been prepared for only the
FOREST model.
$EXIT
:BEGIN
$IF @%%2 = @ THEN SET 2=1
$ANALYZE STRING D=D%%2
INCLUDE BLKSAVE
INCLUDE 'D
* THE INCLUDE FILE MUST GOTO EXIT FOR A NORMAL RETURN;  ELSE,
* FALL THRU TO ERROR STATEMENT BEFORE EXIT.
:ERR
 Sorry, there seems to be an error in the model syntax.
:EXIT
* RESTORE BLOCKS, IF THERE WERE ANY
$IF %%BLOCKS > 0 THEN ANALYZE BL LOAD
$EXIT
```

Figure 12. The Rule File, DISTRIB.RUL

The variations we want to test are in INCLUDE files, D1.INC, D2.INC and D3.INC, respectively. The error switch is turned off before testing the string variable, D. If the string variable does not exist, an error condition is reached, but the message is suppressed by having turned off the error switch. In that case, the command ANALYZE STRING D=1 sets this string variable, which we can think of as a default. In either case, we reach the label OK.

The first command after OK turns the error switch back on. Then, it saves whatever blocks may exist in order to erase them for use during this rule instantiation. The BLKSAVE.INC file is shown in figure 13 (it can be included in other rule files).

The parameter, BLOCKS, is defined to record the total number of blocks (rows + columns). Then, if there are any blocks, they are saved before they are erased.

```
* THIS SAVES BLOCKS, IF ANY
$SET BLOCKS = %NRBLKS
$CALC BLOCKS = BLOCKS + %NCBLKS
$IF %%BLOCKS > 0 THEN ANALYZE BLOCK SAVE
$ANALYZE BLOCK ROW ERASE *
$ANALYZE BLOCK COL ERASE *
```

Figure 13. The Include File, BLKSAVE.INC

After the blocks are erased, the rule file has the statement, INCLUDE D'D. The value of string variable D replaces 'D in this statement, so the variation is D1, D2 or D3, according to whether the value of D is 1, 2 or 3, respectively. As indicated by the comment, each variation contains a branch to the EXIT label unless there is an error. In any case, processing reaches the statement that loads the original blocks, if there were any, before terminating with the EXIT command.

To discuss what these variations do, begin by reading in the FOREST packed file and syntax file. This is done in figure 14 with EXEC FOREST. Then, DISTRIB is passed to the TEST rule file in order to instantiate the DISTRIB.RUL rule file.

```
ANALYZE ... EXEC FOREST
 (messages not shown)
ANALYZE ... I TEST DISTRIB
```

Figure 14. Begin Test of Distribution Rule (DISTRIB.RUL) with FOREST

In figure 15 the first variation is shown. Since no parameter was passed after DISTRIB, the default is used, namely D1. The DISTRIB rule file includes D1.INC. The user sees the response shown.

```
Examine the following block picture.
                BLOCK PICTURE
          H H H
          R R R
          V V V
          S S S
          T T T
          . . .
          B G N
          B A R
DEMAND.1      +
DEMAND.2    +
DEMAND.3    + +
DEMAND.4      +
DEMAND.5 +    +
The row blocks correspond to demands in each of the 5 time periods.  The
column blocks correspond to Harvest activities in the areas that fulfill
the demands. Their meanings are as follows.
   BB is Black Butte
   GA is Glenn Acres
   NR is North Ridge

 (Absence of an area in the column blocks means it does not get
harvested.)
```

Figure 15. Instantiating the First Distribution Rule

Figure 16 shows the first portion of the rule file variation. It begins by setting the submatrix to contain demand rows and harvesting columns having positive level (where $X < .1$ is presumed to mean $X = 0$). It then forms row blocks for time periods, namely D1, D2, D3, D4 and D5 (horizon = 5 in this case). At least one demand block exists unless the FOREST model instance was incorrectly generated. In that case, there is a branch to the ERR label (c.f., figure 12).

Column blocks are defined over the areas of the harvest activities, and at least one must exist for the FOREST instance to be correct. Once we get through these tests, the value of parameter N is set equal to the number of row blocks, which is the horizon of the instance.

```
$ANALYZE BLOCK ROW ERASE *
$ANALYZE BLOCK COL ERASE *
$ANALYZE SUB ROW D
$ANALYZE SUB COL H X=.1/*
$ANALYZE BL R S D?
$IF %NRBLKS = 0 THEN GOTO ERR
$ANALYZE BL C S H??
$IF %NCBLKS = 0 THEN GOTO ERR

$SET N=%NRBLKS
$LOOP
:NEXTBLK
$ANALYZE BL R REN SYN.%%N DEMAND.%%N
$CALC N = N - 1
$IF %%N > 0 THEN GOTO NEXTBLK
$ENDLOOP
```

Figure 16. First Portion of Include File, D1.INC

The subsequent LOOP command is actually over the blocks. Inside a loop, a label can appear, as NEXTBLK in this case. Then, each row block is renamed from the standard naming by the BLOCK ROW SYNTAX specification that created them. For this case, the row block names are SYN.1, SYN.2, SYN.3, SYN.4, and SYN.5. These are renamed to be DEMAND.1, DEMAND.2, DEMAND.3, DEMAND.4, and DEMAND.5, respectively. Note that the CALC command decrements the parameter N, and the loop ends when the value of N becomes 0. As long as N > 0, the branch is GOTO NEXTBLK to rename the next row block.

This brings us to figure 17, where stack 1 is cleared. The next sequence of commands queue each area for which the harvest block contains a column. This is determined by the LOOKUP command, where VLOOK is set equal to the number of columns in column block, SYN.ar, where ar is the area. This method explicitly goes through each area, which presumes they do not change from one instance of FOREST to another.

The QUEUE command is used, rather than PUSH, in order to have their translations in the same order as appears in the block picture (c.f., figure 15). In addition to queueing the areas, each harvest block is renamed from SYN.ar to HRVST.ar, where ar = area.

```
$STACK 1
* STACK AREAS AND RENAME HARVEST ACTIVITY BLOCKS FOR
* THOSE THAT HAVE POSITIVE LEVELS
$LOOKUP BLOCK COL SYN.BB
$IF %VLOOK = 0 THEN SKIP 2
$QUEUE 1 BB
$ANALYZE BL C REN SYN.BB HRVST.BB
$LOOKUP BLOCK COL SYN.BR
$IF %VLOOK = 0 THEN SKIP 2
$QUEUE 1 BR
$ANALYZE BL C REN SYN.BR HRVST.BR
$LOOKUP BLOCK COL SYN.GA
$IF %VLOOK = 0 THEN SKIP 2
$QUEUE 1 GA
$ANALYZE BL C REN SYN.GA HRVST.GA
$LOOKUP BLOCK COL SYN.NR
$IF %VLOOK = 0 THEN SKIP 2
$QUEUE 1 NR
$ANALYZE BL C REN SYN.NR HRVST.NR
$LOOKUP BLOCK COL SYN.SM
$IF %VLOOK = 0 THEN SKIP 2
$QUEUE 1 SM
$ANALYZE BL C REN SYN.SM HRVST.SM
```

Figure 17. Second Portion of Include File, D1.INC

In figure 18 the first sentence of the response text appears. Then, ANALYZE BL PIC generates the block picture. Note in figure 15 that the block picture has the renamed blocks. After the block picture, the text continues, where the horizon replaces the keyword reference, %NRBLKS.

The subsequent loop goes through the stack, giving the translation of each area. The area code appears as a suffix in the harvest column block, and the ENTITY command translates it. The test before the ENDLOOP branches back to label NEXTAR if stack 1 is not empty (each POP removes an area from the stack).

After the loop ends, the final sentence is put into the response text, and the rule file terminates with the branch to EXIT.

```
 Examine the following block picture.
$ANALYZE BL PIC
  The row blocks correspond to demands in each of the %NRBLKS
  time periods.
  The column blocks correspond to Harvest activities in the areas
  that fulfill the demands.  Their meanings are as follows.
$TEXT 1 MARGIN=3
* Use LOOP to go thru stack to interpret areas.
$LOOP
:NEXTAR
$POP 1 AR
 %%AR is
$ENTITY %%AR AR
$TEXT
$IF %NSTACK1 > 0 THEN GOTO NEXTAR
$ENDLOOP
$TEXT 1 MARGIN=1
 (Absence of an area in the column blocks means it does not get
 harvested.)
$GOTO EXIT
```

Figure 18. Last Portion of Include File, D1.INC

Figure 19 instantiates the second variation by passing 2 as the second parameter in the INTERPRT specification. This variation is more succinct and does not present a block picture to the user. It simply goes through each area for which harvesting occurs in some time period, and gives a sentence that says which time periods each area has positive levels of harvesting.

```
ANALYZE ... I T DISTRIB 2
 The harvesting is distributed over the 5 time periods as follows.
 Area Black Butte is harvested in 5.
 Area Glenn Acres is harvested in 2, 3.
 Area North Ridge is harvested in 1, 3, 4, 5.
```

Figure 19. Instantiating the Second Distribution Rule

Figure 20 shows the include file that produced this variation. It begins by setting the submatrix as in the first variation. The first sentence is obtained from the text template with the horizon replacing the keyword reference, %NROWS. Since the rows in the submatrix are precisely all demand rows, this number is the number of time periods, which is the horizon. The template does not have a period, but this gets added with the logic that follows.

```
$ANALYZE SUB COL H X=.1/*
$ANALYZE SUB ROW D
$ANALYZE SUB SET

$SET AREA=
 The harvesting is distributed over the %NROWS time periods as follows
$LOOP
$SET AR %COLUMN(2:3)
$SET T  %COLUMN(4:4)
$IF @%%AR <> @%%AREA THEN GOTO NEWAREA
, %%T
$SKIP LOOP
:NEWAREA
.
$TEXT
 Area
$ENTITY %%AR AR
is harvested in %%T
$SET AREA=%%AR
$NEXT COL
.
$GOTO EXIT
```

Figure 20. Include File, D2.INC

For the first entry in the loop, the value of parameter AREA is blank. Therefore, the test with %%AR branches to NEWAREA, which adds a period, then it clears the response text. This is how the first sentence was formed. New text begins with 'Area' followed by a translation of the first area, which is the value of parameter AR. The time period in which this harvesting occurs is the value of parameter, T, which was set equal to the character, %COLUMN(4:4). This follows the syntax of the FOREST model, where the fourth (and last) character in the name of a harvest activity is the time period.

Before reaching the NEXT command, the value of parameter AREA is set equal to the value of parameter AR. Then, when we advance to the next column and go back to the top of the loop, we continue with additional time periods if the next column harvests the same area (the names are stored in sort order, so if the area changes, we are finished with that area).

Note how the commas are inserted for areas that are harvested in more than one time period. If the values of parameters AR and AREA are equal, the text template is ', %%T', and %%T is replaced with the time period of the column name. Then, the SKIP LOOP goes to the next column. If the area changes, the branch to NEWAREA inserts a period to complete the sentence.

Figure 21 is the third variation. The response is a listing of the demand equations with the harvest activities having positive level. Following the equation listing, each harvest activity is explained.

```
ANALYZE ... I T DISTRIB 3
 The following demand equations show the harvesting activities that
were used.
   1050   <= D1      =   0.4 HNR1
   1100   <= D2      =   0.3 HGA2
   1150   <= D3      =   0.3 HGA3 +  0.4 HNR3
   1200   <= D4      =   0.4 HNR4
   1250   <= D5      =   0.3 HBB5 +  0.4 HNR5

   Column HBB5 harvests timber in Black Butte in period 5.
   Column HGA2 harvests timber in Glenn Acres in period 2.
   Column HGA3 harvests timber in Glenn Acres in period 3.
   Column HNR1 harvests timber in North Ridge in period 1.
   Column HNR3 harvests timber in North Ridge in period 3.
   Column HNR4 harvests timber in North Ridge in period 4.
   Column HNR5 harvests timber in North Ridge in period 5.
```

Figure 21. Instantiating the Third Distribution Rule

Figure 22 shows the include file that produced this variation. It simply gives the text literally, followed by the ANALYZE LIST command. After skipping a line with the TEXT command, the ANALYZE EXPLAIN produces the column explanations.

```
The following demand equations show the harvesting activities
that were used.

$ANALYZE LIST D,H X=.1/*
$TEXT
$ANALYZE EXPLAIN COL
$EXIT
```

Figure 22. Include File, D3.INC

You might also examine INFEAS.RUL, which uses the same technique to test rules for diagnosing infeasibility.

Chapter 12. Using The Modules

Man possesses powers of various sorts which he habitually fails to use.
Po L. Yu

This chapter applies only to the full version, where source codes are included (contact the author for information about the availability of the full version). If you have the full version, you received two 3.5" disks. All ANALYZE source codes, including the FLIP, VIEWS and GETMAT modules, are on one disk. The UTILITY disk has source codes for all utilities, including LPRENAME.

Figure 12-1 shows the calling sequence among the modules. The main program, which interacts with the user, is in the FLIP module. It performs routine functions, like parsing and spooling, and calls upon the routines in ANALSYS for initialization and the interactive session. The VIEWS module supports the SCHEMA and GRAPH commands, and GETMAT is the matrix/solution interface to support procedures and query. The sections in this chapter describe these modules, and the source code listing are fully annotated to provide system documentation (see, also, documentation files, ANALDATA.DOC and GETMAT.DOC).

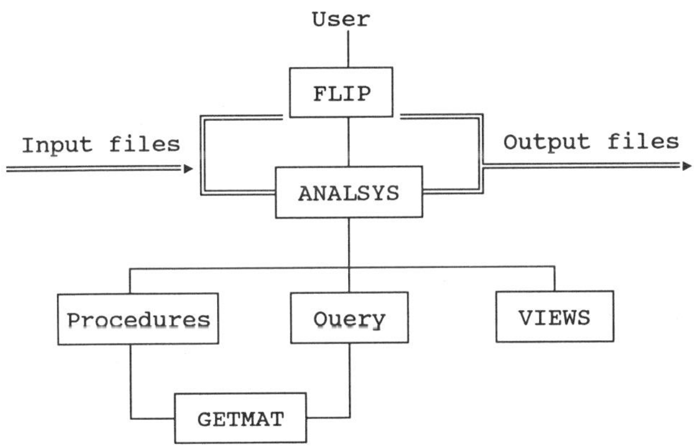

Figure 12-1. Calling Sequence of ANALYZE Modules

The next section describes how to prepare the source codes for an alternative computing environment. The remaining sections describe the support modules, FLIP, GETMAT and VIEWS.

12.1 Setup for an alternative computing environment

Besides DOS, ANALYZE has been tested on a PRIME 9950/PRIMOS, VAX 8700/VMS, IBM/CMS, IBM/TSO, and SUN4/UNIX. It is written in FORTRAN/77, so in principle it runs in any environment, but there are at least 2 things to consider. One is the file protocol, the other is the INCLUDE syntax for the compiler. For those reasons, and for increasing the dimensions to accommodate larger problems, the source codes must be edited before uploading them to a mainframe.

It is recommended you install ANALYZE on your microcomputer, even if you intend to run in another computing environment. Go through the suggested exercises to become familiar with ANALYZE before uploading.

The preparation is automated in SETANAL.BAT, but you may need to modify some things if your environment is not explicitly recognized, such as to prepare for use of extended memory on your microcomputer (either for a Lahey compiler or Microsoft(R) under OS2). To use the automatic preparation, create a directory on your hard disk and make sure it is empty (so you can upload all the files that will appear).

Insert the 3.5" ANALYZE disk in your a: drive. At the DOS prompt, enter:

```
> A:PKXARC  A:ANALSUBS  SETANAL.BAT
```

Then, simply enter:

```
> SETANAL  environment [drive]
```

where *environment* is one of: XT, VAX, TSO, CMS, XNX, AIX, SUN4, PRIME. This must be given in CAPS. This will extract source and data files for upload. The file MSFILES.LNK will also appear. This shows the files to link, with Microsoft(R) overlays indicated by the parentheses. The main program is in FLIPMAIN, and you may want to make a library out of the GETMAT source codes.

The default drive is a, but you can specify the drive that contains the ANALYZE disk. For example, to setup ANALYZE for a VAX environment from drive b, enter SETRAND VAX B (note no colon after the drive).

<u>Note to VM/CMS users</u>
See NOTES.CMS for notes about upload, compilation and linking, and see INSTALL.CMS for easy installation.

<u>Note to VAX/VMS users</u>
See NOTES.VAX, and see other *.VAX files for command files to assist installation.

You might want to change the dimensions, especially if your computing environment has more memory to accommodate larger problems.

The INCLUDE files, DCANAL and DCGETMAT, have PARAMETER statements that can be changed before compiling. The AUTOEDIT, used by SETANAL, does this, and you can see the settings for each environment. You may edit these yourself to make further changes (but keep a backup of the original).

To estimate the storage, GETDATA.DOC gives details and examples. Briefly, there are three primary arrays, which are declared in DCGETMAT:

NAMERC.......... character array to hold names of rows and columns;
INDEX... .... integer array to hold pointers and indexes;
VALUE. .... real array to hold distinct coefficients and solution values.

The following problem statistics govern the space needed.

m = number of rows
n = number of columns
z = number of nonzero coefficients
v = number of distinct coefficient values (magnitudes only)
L = number of characters in row and column names

Note: v includes bounds and right-hand sides.

The NAMERC array size is simply $L \times (m + n + 2)$. The lengths of the INDEX and VALUE arrays are governed by the numeric dimensions (m, n, z, v). Besides just storing the LP matrix and solution, INDEX and VALUE require additional storage for the basis (to use the BASIS and RATEOF commands). Additional VALUE storage is needed for the REDUCE command. Rules of thumb to estimate storage requirements are given in GETDATA.DOC, which can be reduced to just the maximum numbers of rows and columns.

You must also change the PARAMETER specifications in DCANAL, which contains a dictionary of what the parameters mean. It is especially important that you set the maximum numbers of rows and columns to agree with the above.

The LPRENAME utility can be prepared in a similar manner, using SETLPRN to prepare the source codes for the target environment. HPAGER and FORTREF are designed only for the DOS environment.

The rest of this chapter describes how you can use the primary modules that support ANALYZE. These can be linked to any applications program written in FORTRAN/77. The first module is a *FORTRAN Language Interactive Processor* (FLIP). It contains a main program to control an interactive session plus a library of subroutines for parsing and other routine, application-independent functions. The second module, called GETMAT, manages a database composed of a linear programming matrix and (optionally) solution. The third module, called VIEWS, manages displays of schema and graphs.

You can use FORTREF to obtain a cross reference listing of your system. This was done for ANALYZE, and it is contained in the FORTREF.ARC file. If you installed FORTREF, you can print ANALYZE.REF to see the (sorted) list of subroutines, including FLIP, GETMAT and VIEWS, and their calling sequences. The source code listing for each module begins with a dictionary of subroutines it contains, and each subroutine has comments for details.

12.2 FLIP

The *FORTRAN Language Interactive Processor* (FLIP) is a main program plus a collection of subroutines to perform routine functions associated with an interactive session.

By itself, FLIP has three unresolved external references: SYSIN, SYSCMD and SYSDBG. These comprise the links with the application system, such as ANALYZE. The SYSIN subroutine is called before the interactive session begins. It returns to FLIP information about the commands, switches and other information that FLIP uses to parse user specifications. It is also the place for initializing portions of the application system. The SYSCMD subroutine is called by FLIP when the user specifies a *procedural command* - that is, one of the application system's commands (as opposed to the FLIP intrinsic commands, called session control commands). The application system finishes the parsing, using the FLIP subroutine library. The SYSDBG subroutine is called by FLIP whenever the user enters a comma, which is a hidden command, designed to help debugging.

These three subroutines for ANALYZE are in ANALSYS.FOR. Examine the structure for your own applications. Note that SYSCMD is mainly a computed GOTO statement, branching on the command number passed by FLIP. Executions of those commands are done by other subroutines.

The FLIP subroutine library is divided into nine files, shown in table 12-1. Additionally, FLIPSYS.FOR contains SYSIN, SYSCMD, and SYSDBG subroutines to run FLIP by itself. Although this is primarily for debugging FLIP, it is included in the system (it can be used for keyword capability, such as generating SPEC files for OB1 or MINOS).

FLDICTRY............dictionary support
FLFILES..................file specs
FLIPCMS ....CMS special routines (primarily I/O)
FLIPMGT ....initial sign-on and debug
FLKEYWRD..........keyword support
FLPARSE................parsing
FLRAND.................random number generation
FLSCREEN ....screen management
FLSTRING ....string management

Table 12-1. FLIP Source Files

The SYS files can be structured as follows. File FLIPMGT.FOR contains SUBROUTINE FLINIT, which is called once by the main program, in file FLIPMAIN.FOR, to initialize the session. This subroutine calls SYSIN with the following arguments.

SYSNAM.................System name (used at prompt)
VERSN ....Version (used at sign-on)
NCSYS ....Number of commands
COMAND ....Array of commands
NSWSYS..................Number of switches
SWNAME ....Array of switch names
SWVALU ....Array of switch values

The system name (SYSNAM) is 1 to 8 characters, and it is used as a default file name for keyworded files, like the dictionary and keywords. The version (VERSN) can be 1 to 64 characters. ANALYZE uses only 4 characters for its version passed to FLIP.

The number of commands (NCSYS) upon entrance into SYSIN is set to the maximum allowable (net of the FLIP intrinsic commands, called session control).

The SYSIN routine must set this equal to the number of procedural commands, which is the length of COMAND upon return. (FLIP merges its intrinsic commands with the ones declared in COMAND(1), ..., COMAND(NCSYS) returned by SYSIN. It is allowed to set NCSYS=0, in which case only the FLIP intrinsic commands are available to the session. This is useful, for example, just to have the keyword capability, such as for generating a spec file for OB1 or MINOS.

The number of switches (NSWSYS) upon entrance into SYSIN is set to the maximum allowable (net of FLIP switches). The SYSIN routine must set this equal to the number of additional switches. Each switch has a name (SWNAME) of 1 to 8 characters, and an initial value (SWVALU) of either true or false (LOGICAL*1). It is allowed to set NSWSYS=0, so only the FLIP switches are available to the session.

After setting what FLIP needs, the SYSIN routine may initialize its own information. See ANALSYS.FOR for an example of how this is done.

The SYSCMD routine has the following arguments.

CMD........................Procedural command number
SWVALU................Switch values
NSWSYS.................Number of switches
CLIST.......................User input (character list)
FIRST.......................First parse position of CLIST
LAST........................Last parse position of CLIST
RCODE...................Return code

The command number (CMD) corresponds to COMAND(CMD), set in SYSIN. All switch values (SWVALU) are passed, with the total number (NSWSYS). The switches set in SYSIN must use an offset in SYSCMD.

The user input that remains to be parsed is CLIST(FIRST:LAST), where CLIST is CHARACTER*128. It must be that FIRST > 1, because the first specification is the command, which FLIP already parsed to obtain the command number (CMD). If FIRST > LAST, there are no more characters to parse.

The return code (RCODE) tells FLIP if all went well. If so, SYSCMD (or one of its subroutines called) sets RCODE=0. Not changing RCODE is the same because FLIP sets RCODE=0 before calling SYSCMD. A negative return code causes FLIP to print a syntax error message, but it does not abort an execution file. A positive return code causes FLIP to print a fatal error message, and, unless the error switch has been turned off (i.e., set False by the user), an execution file is terminated. In addition, in that case, output also returns to the terminal if there was an output file active.

The file ANALSYS.FOR should be examined for a possible structure of the SYSCMD routine. After using the switches for updates of settings (such as the message switch, which is intrinsic in FLIP), it is primarily a computed GOTO statement that branches on the command number. Other routines actually execute the commands, including additional parsing using the FLIP subroutine library (notably, FLPARSE.FOR).

For example, FLPARSE.FOR has the following subroutines to assist parsing using specifications.

FTOKEN ......Gets token from CLIST(FIRST:LAST)
FOPTN ....Gets option from CLIST(FIRST:LAST)
FOPSET ....Gets set of options from CLIST(FIRST:LAST)
FLOOK ....Looks for string in CLIST(FIRST:LAST)
FVRNG.........Gets range from CLIST(FIRST:LAST)
FMATCH......Compares 2 strings
FC2R..............Converts character to real
FR2C..............Converts real to character*8
FR2C12 ....Converts real to character*12
FI2C ....Converts integer to character
FC2I ....Converts character to integer
FSQUEZ.......Squeezes blanks out of string

You can begin documenting your applications system with the files that have suffix .FLP. For example, INTRO.FLP is the first part of INTRO.HLP in ANALYZE, and SETUP.FLP is the basis for SETUP.DOC.

12.3 GETMAT

The GETMAT module manages access to matrix and solution information for an LP. It is self-contained with all array declarations in DCGETMAT. This is deliberately in conflict with DCANAL to keep this module separate. The data documentation is in GETDATA.DOC, and the source listings contain much of the code documentation.

The basic functions are storage and retrieval of LP information in the handle, rim and body (see chapter 2). The GETMAT module also executes some of the algorithms, such as successive bound reduction, because of its efficiency at the low level of information access. These subroutines, like those in the file

GREDUCE.FOR, could be replaced with a dummy routine if the algorithm is not needed for another application.

The GETMAT subroutine library is divided into 14 files, shown in table 12-2.

GBASIS.............. executes basis routines
GBRANGE........ computes ranges
GETEDIT........... revises bounds
GETIOMAT....... reads matrix file
GETIOPCK........ reads/writes packed files
GETIOSOL........ reads solution files
GETIOSUB........ writes submatrix files
GETLIB.............. access to LP information
GETRANS......... executes transformations
GETSET............. initializes and manages submatrix maps
GREDBIN.......... executes logical portion of reduce algorithm
GREDEXP......... executes extended reduce algorithm
GREDMAT........ writes reduced matrix file
GREDUCE........ executes reduce algorithm

Table 12-2. GETMAT Source Files

You must CALL GINITL first (in your SYSIN routine if you are using FLIP) in order for GETMAT to initialize itself. This is in file GETSET.FOR, which also contains other routines for setting LP information, including the management of row and column maps (as used for submatrix representation in ANALYZE).

Each of the GETIO source files are self-explanatory, and you might use only some of them. (For example, LPRENAME does not use GETIOSUB.) The primary source file for access to LP information is GETLIB.FOR. This is used extensively by ANALYZE and by other GETMAT routines. Here is a list of subroutines it contains.

NAME................. Gets name from number
GPTNAM........... Puts name
GETNUM........... Gets number from name
GETNAM........... Gets name from number
GETBND............ Gets bound values
GETRIM............. Gets rim values
GETYPE............. Gets type of row (MPS format)
GETSOL............. Gets solution values
GETST................ Gets solution status
GETFNM............ Gets first name match
GETCOL............ Gets column nonzeroes

GETROW........... Gets row nozeroes
GETSGN .... Gets sign of nonzero
GETVID .... Gets value index (adds to pool)
GETVAL............ Gets value from index
GETBVP............. Gets basic variable from pivot position
GETPIV .... Gets pivot position from basic variable
GPRDCT .... Gets inner product

The LP matrix is stored by columns with a (variable-length) list of nonzero row and value indexes per column. It is, therefore, not advisable to CALL GETROW for access by rows. This works all right in ANALYZE for query purposes, such as to support the SHOW ROW specification, but it is inefficient for other uses, like optimization.

12.4 VIEWS

The VIEWS module contains the schema management, which can be used by other programs, and routines for the graph view, which is bound to ANALYZE.. The only source file for the schema view is SCHEMA.FOR, which has the following subroutines.

SCHCLR .... clear schema
SCHPUT .... put entry into schema
SCHDSP.............. display schema
SCHWRP .... write packed schema (unformatted)
SCHRDP............. read packed schema (unformatted)
SCHFIL............... open schema file
SCHSKP.............. skip comment lines on DATFIL
SCHDBG .... debug

These routines are simple management functions. They do not create schema by themselves. The host system is responsible for that. For example, ANALYZE has the source code ASCHEMA.FOR (other systems, like MODLER, have their own linking routines).

A schema is simply a table whose entries are symbolic. It is intended for small numbers of rows and columns, so it is stored in full matrix form. If a schema display is requested, the schema module decides if it can fit rectangularly (without horizontal scrolling). If not, it is displayed in relational format.

Subroutines SCREEN.FOR and SCRPLOTS.FOR support the plot options with the BLOCK and SHOW commands. These can be used for other applications.

Subroutines GRAPHDO.FOR, GRAPHIO.FOR and GRAPHSYN.FOR support the GRAPH command in ANALYZE. Unlike the schema and screen modules, the graph subroutines rely on DCANAL and cannot be easily detached for other uses. The same is true for TABLE.FOR, which supports the TABLE command and is tied to ANALYZE.

Appendix A: Error Messages

This appendix describes error messages. The print file, ERRORS.DOC, begins with the following.

```
Error messages are given when ANALYZE cannot recognize what you
mean, or if some necessary condition is violated.  Syntax errors
are the most common cause, and I/O errors are the next most
common.  As in debugging a linear program, one takes information
and tries to diagnose the cause.  Because error messages are
imperfect in the information they provide, this list of error
message explanations is not a complete guide to debugging.
Further, ANALYZE is an imperfect language with some limitations
and quirks.
```

This file is your best starting point for understanding how to use the information from an error message.

Besides syntax, another common cause of an error is the I/O ERROR message. This occurs if the file does not exist. Check your full file specification by using _SETUP to see the default prefixes and suffixes. You might also have inadvertently put an unprintable character in the name of the file, or in the NAME specification in the matrix file (or in the modeling system that generated it). Then, when you try to read data using this name as the default, you get an I/O error message. Try over-riding the default with a fully specified data file. If that works, you probably have an error in the name, in which case you need to go back to your editor and re-enter the name.

Appendix B: The ANALYZE Library

Most of the linear programs that were installed with ANALYZE are for testing the commands and providing a few interesting exercises, which were covered in this User's Guide. There is a second library, called MOREDATA, with more instances of linear programs that can be used for similar exercises.

See MOREDATA.DOC for information about this library. Then, if you want to add them to your ANALYZE system, put the ANALYZE DATA disk into your drive and enter A:PKXARC -R A:MOREDATA (change the drive prefix if your floppy is not in drive A). Then, enter ANALYZE (Standard version) and enter EXEC MOREDATA. This will read each matrix and solution file and write a packed file, so your subsequent analysis can read the packed file more quickly (and the Views version can be used).

Index